Pigs, Poultry and Poo

An Urbanite Couple's Journey to Country Life

JASON GIBBS

THE CROWOOD PRESS

First published in 2012 by
The Crowood Press Ltd
Ramsbury, Marlborough
Wiltshire SN8 2HR

www.crowood.com

British Library Cataloguing-in-Publication Data
A catalogue record for this book is available from the British Library.

ISBN 978 1 84797 391 7

Frontispiece: our alpacas, Algy and Verdigris, looking inquisitive.

Typeset by SR Nova Pvt Ltd., Bangalore, India

Printed and bound in India by Replika Press Pvt Ltd

Contents

Introduction

This is not a story about two people leaving London to make a new home in the country, although that does occur. It is also not a story about converting a barn, in part because we haven't finished. This is a story about animals and how we've come to care for an eclectic menagerie. Our aim is not to teach – though hopefully our mistakes and missteps will be useful to anyone considering taking on the responsibility of animals – but to entertain, and perhaps counsel caution.

The Plan

My wife and I used to live in a lower ground floor flat in Islington (or a basement flat in Hackney, depending on who's describing it... I maintain that we paid Islington council tax so therefore it was in Islington – but I digress). I worked in IT in the City, a thirty-minute bus ride, or forty-minute walk away (literally), and Alex (my wife) worked at a law firm in Westminster, a rambling sixty minutes away each morning. We had a small outside space, about ten feet by ten feet, which was cut out of a larger garden owned by the flat above us, but it was enough for the occasional barbecue, and for my smoking friends to hang out in when the urge took them. We were often out at local restaurants, or exploring the nicer end of the food industry around the rest of London, and we could go out with friends at the drop of a hat. It was, all in all, a good life (though obviously not 'The Good Life') – so why did we leave all this and move to a dilapidated and unconverted barn in Hampshire? What madness overtook us?

Firstly I'd say that we weren't upset by the hollow consumerism of our lives: we did indeed love living in London, and enjoyed the facilities of a large metropolis. It was more that when it came down to it we didn't see ourselves living in Islington forever, and having pets or children there would be more challenging than perhaps we wanted. So we started thinking about moving out. We quickly realized that we weren't entirely keen on the suburbs, because even though there was more space, it was still fairly packed, and of course house prices were somewhat discouraging.

Alex was from the wilds of Hampshire, and in my peripatetic youth I'd lived in some more rural locations, and gradually the siren call of the countryside began to impinge upon us. One concern I had about moving to the country was whether I had become too citified and would not be able to cope with the fresh air, greenery, animals, bugs and muck, but we experimented with a few trips outside the M25 and nothing tried to eat me or cover me in poo, so I felt a bit more confident. To be fair, my history with creepy crawlies at this point was not impressive, as Alex often reminds me... One night we were sound asleep in the flat and a giant spider landed on my face, waking me up and completely freaking me out – so much so that I insisted that we move to the spare room. Spiders in the distance are OK, but on my face, that's horrible! Alex also wants me to point out that it was unlikely to be a 'giant', it was probably just a normal-sized spider – but then how would she know, she was fast asleep at the time! This incident might have held us back, but Alex usefully told me to 'toughen up'; so with that resolved, we were all set...

Fish Keeping

Another reason we started looking in the countryside was that I'd decided I wanted to be a fish farmer. (This would, of course, have made Alex a fish wife, which still brings me amusement to this day...) I read an article in *The Economist* about the 'Blue Revolution' and how we'd be able to feed the world with fish farmed using recently developed techniques. It had everything: new technology, food, fish and an underlying mission. Some people might question whether I should have been looking for a loch or bit of Scottish coast, but I had a very specific plan. I was going to farm tilapia, a white fish which is vegetarian (in the sense of eating vegetable matter, not that it is a veggie option on menus), which to my mind is rather important, in that each kilogram of farmed salmon or trout has been fed on 4kg of wild fish caught from the sea (at least this was the case in 2003), which didn't sound particularly sustainable to me.

I researched the topic at some length, covering integrated farming – a simple example being chicken poo fed to fish, fish poo fed to grains, grain fed to chickens, with the farmer taking surplus chicken, eggs, fish and grain out of the system – and subscribing to fish farming magazines. In the end I decided to try and develop a system similar to one I'd read about in the Lebanon, where a series of pools were connected with pipes so that as the fish matured they could be moved between pools until they could be taken out for food. I also had a crazy idea of letting the tilapia swim in a mixture of water and white wine for their final week or so to give them some extra flavour...

So there I was with a plan, some rough costings, and the requirement for some outbuildings. Alex was surprisingly supportive during this time – usually she spots when I've developed some hare-brained scheme and taunts me mercilessly until I give it up. I suspected it was because she felt it helped us to move out of the Big Smoke. Actually she had an alternative strategy ... she bought me some fish for Christmas.

Not just any fish: some tropical fish. Well, actually she bought me a large tank, and then later we both went to get the fish together. The tank was a rectangular fifty-litre glass box, with all the filters and heater in one corner, and was specifically designed for tropical fish. The instructions were to set it up, get some water in it and add a few chemicals, and then leave it for a week or so to allow it to settle before putting in any fish. I was so excited, the anticipation was excruciating, and I believe is part of a conspiracy between tank makers and fish sellers to encourage people to buy more.

After a breathless week of waiting we headed to the Sea Dragon, our local live fish vending establishment, and with a bit of advice, took home some tropical fish. As I recall we bought two albino tiger barbs (which we've since learned are aggressive fish and probably not good to start off with), four zebra danios (small, thin-striped fish), four neon tetras and four harlequins (which for some reason I named after the turtles, Leonardo, Michelangelo, Donatello and Raphael).

The arrival of the fish revealed a couple of traits in me which I'd kept long hidden: I can be slightly obsessive, and a little daft. I bought several books on keeping tropical fish, and a log book (in fact a normal lined paper book, but I used it as a log book). In the log book I recorded each time I fed the fish, each time we cleaned out the tank, and, sadly, each time a fish died – I'd even draw a little tombstone with RIP written on it. This lasted for about four months until I relaxed somewhat about the fish, and realized that my log was telling me very little. The books revealed that we knew very little about tropical fish: to summarize – they are very sensitive to temperature variations, and they can die at the drop of a hat. One book recommended we had our fish autopsied when they died, but that seemed somewhat ridiculous, and if possible, I'd guess rather expensive.

There are lots of diseases that affect tropical fish, and not many treatments, but one thing was clear from the books – stress was bad for them. I was considering arranging massages, and perhaps calming trips to a spa . . . but instead we were very careful with the water changing, and we added a chemical called something like 'Stress Relief', which apparently helps to chill out the fish, and another chemical which helps to protect them from damage and some of the more common bugs. I think all the zebra danios died within a week or so of each other, but after that the tank settled down. One of the harlequins was often off on its own, clearly Raphael taking after his namesake.

This first experience of keeping animals was illuminating. The main chores were feeding them, and fishing out the occasional dead one, making it a low effort endeavour – except when cleaning out the tank. The authorities differ on what percentage of the water needs to be exchanged, but we were blessed with a system which did not require a full change of the tank, which I can imagine would be a complete nightmare. Instead we used to replace about a third of the water each time. The first part was obviously to get water out, which initially we accomplished using a pint glass. This was not very efficient, and also didn't allow us to get to the muck at the bottom of the tank, so we bought a siphon kit. The siphon had a small hand pump on it and in theory would, once the water started to flow, allow us to use the end to suck up all the debris around the tank; it also had a filter to prevent any fish from being caught. Unfortunately the pump was rather inefficient, and the only way to get the siphon working was for me to suck on the output end until the water flowed out. About half the time I'd be able to get the water out without any ending up in my mouth – the other half of the time it was pretty unpleasant!

The final touch for the aquarium was to add some plants. We bought a few of the standard tropical plants and they added a lot of colour to the tank, and also gave the fish more places to hide, which hopefully helped de-stress them further. Something else also arrived with the plants: snails, probably carried as eggs on the plants. We were more than happy with these additional denizens of our watery pet area. I'd been quite tempted by getting a plastic skull, or maybe a castle, but they were just a little too tacky, and at least one of my fish books had said they were potentially poisonous to the fish due to the paints used in them. This didn't sound entirely logical in that they were made specifically for fish tanks – but better safe than sorry.

So eventually the fish were all settled, we had a lovely aquarium in the corner of the room, and Alex thought that would be the end of my ramblings about becoming a fish farmer. She was wrong!

Looking for a Property

In the autumn of 2003 we started looking for a place to move to. Not in a very active sense, this was very much about getting a feel for the market, so we searched a plethora of websites. We hadn't quite settled on a place to move to, but I was very clear that I wanted a commute of less than ninety minutes, and preferably sixty. I was working next to Liverpool Street, so this meant we needed to look north-east of London, but that it should be a simple train ride in. We started to concentrate on Suffolk, for a number of reasons, but in essence we didn't want to live in Essex or Hertfordshire, as they were both too close to London, too expensive, and didn't offer the types of property we wanted. Norfolk and Cambridgeshire were too far and too flat (according to Alex) respectively, and so Suffolk stood out clearly. We weren't completely bound to the one county and searched for places in northern Essex, and Hertfordshire, and southern Cambridgeshire as well. Regarding the type of property, we decided to look for a place that was somewhat run down: not only would this help bring the price down, but we could then add some value by fixing it up – and Alex's parents, Gordon and Sue, had expressed some willingness to help us.

Now Suffolk is a lovely county, mostly countryside, though to some tastes it still lacks a certain hilliness. We actually viewed a place north of Stowmarket, an old rectory right in the middle of nine acres of fields, with eight bedrooms and a flood in the cellar (or indoor pool, as I thought of it!). We did quite like it, but the indicative price was just beyond our limit, and then we were told it was going to go to sealed bids, which was likely to increase the price further. Besides, one of the other couples viewing it had brought their architect, which left us feeling we were out of our league…

We continued to look for several months, and found nothing suitable – it was all either too expensive, too far away in commuter terms, or didn't have the facilities that we (I) wanted. So the search widened to include counties south and west of London too. I was afraid of moving south of the river – after all, I had spent eight years in North London disdaining those who were on the other side! Alex told me not to be silly, and that was that.

Having widened our search area there were more properties available, and we even viewed a couple of places in East Sussex, although this would have been a mighty commute into work. One was a farm with an astonishingly ugly house which faced north and on to the road, but which had a fabulous Sussex barn and collection of outbuildings. We thought it would make a wonderful conversion, and I decided to phone the planning department to find out what we'd need to do. (This is actually a moot point, as I hate using the phone and so Alex nearly always does the phoning. She disputed that I'd made this phone call, but as I could remember what the man had said and she couldn't, she relented … but I think she still doesn't believe me.) It transpired that in order to convert a barn in East Sussex we had to arrange for Hell to freeze over, so that idea was dropped. Also the man in the planning department was grumpy, which reinforced my general aversion to phoning people! However, the idea of doing a barn conversion was now in our heads.

I did see another barn in East Sussex, which my friend Adam kindly drove me to. It had full planning permission, but it was in the middle of a large working farm, with a huge metal barn being built just next to it. Also the planning included the, to me, unforgivable step of putting a first floor into it, effectively just converting it into a normal house within a barn structure. My view is that if you live in a barn it should feel like a barn, and if you want to live in a modern four-bedroom house there are tens of thousands to choose from, whereas there are only a few barns. I was starting to become a bit of a barn zealot. Also I was amazed to learn that we were the third viewing that day, and they'd had five people the day before, which had been the first day of viewings! I therefore guessed it would go for a higher price than we could afford anyway…

Finally in the summer of 2004 we took a look at Hampshire. I'd always said that it was too expensive, but one thing we'd discovered in our nine months or so of searching was that the type of property we were looking for had its own price band, which was almost independent of any local pricing. However this, and a nice pay rise for each of us, meant that we could look at the prices without falling over, and some of them we could even afford!

We were visiting Alex's parents, who still lived in the village she'd grown up in, the weekend before Alex was due to go on a two-week trip to Australia; they mentioned that they'd seen the details of a farm and barn being sold in lots just ten minutes down the road. The pictures on the internet were not entirely encouraging, but we thought we should visit it as we were in the area. Despite a feeling of weariness, off we trudged to look at it. Our first view was not exactly stunning ... there was a large metal and asbestos barn, some ugly blockwork buildings, a partially dismantled Dutch barn, and almost hidden behind all of these, a tin-roofed wood-framed barn connected to some brick buildings. However, once we walked into the barn, it was a different matter ... if you looked past the straw and the suspiciously squidgy brown piles on the floor, and the cobwebs and dust and the large tractor, there was a stunning building trying to get out. It was connected in an L shape to a brick cowshed that had chickens in it, which led on to another wooden barn with a hayloft. This was full of cobwebs and dust, and the hay-loft had holes in the floor big enough to fall through. I was looking around thoughtfully, with my brain moving into the 'well, it would be nice but we probably can't do it' realm, when I saw Alex's face: she was besotted. We were in big trouble!

Aquiring Our Barn

We decamped to a local tea shop to discuss ideas. Gordon was excited too, as this was the kind of big project he really wanted to do. He had been running his own business for several years doing a combination of small building works, painting and general maintenance of property – fixing up windows, replacing doors and suchlike – and while he'd done some interesting work, he hadn't had anything that had really caught his imagination – until now! With Alex and Gordon excitedly chatting through options, I was running the potential numbers through my head, and coming up a bit short, I have to admit. Still, after some discussion we seemed to have a workable plan: we would do a temporary conversion of the cowshed and live in that while we converted the rest of the barn, and we'd take as long as necessary to do it. No caravans for us, and no rushing.

I showed the website to one of my friends at work, and he thought I was joking. Then he thought I'd gone insane, and informed me there'd be an awful lot of work. For a start we'd need to repoint all the internal brickwork ... this was after seeing a picture of one of the internal columns in the barn. This really made me laugh, and when I revealed that repointing was the least of our problems and indicated the true size of the task, he decided I really was crazy – and as was his way, decided to buy some popcorn and sit back and watch, with the occasional acerbic comment to keep things going!

The sale went to sealed bids, and we had a tough time deciding what to bid, but we really wanted it so went for the maximum amount we could possibly afford. In addition we had to agree to the challenging task of completing within four weeks of the offer being accepted, but we lined up our solicitors and the mortgage company, making both aware that this was part of the bid, and they both indicated that they believed it was possible, and when pushed, agreed that they would do their utmost to make it happen, which is about the best one can hope for. This was only a week after we'd seen the place, and as Alex was away I had to put the bid together and send it out without her usual guiding eye (being a solicitor, she is very sharp on such things); when I sent the bid in, I was more worried that I'd messed up the details than that the price was too low. Based on what the estate agent said when he phoned to tell me the good news, I suspect we blew away the opposition, but despite being slightly galling that wasn't important: we now had the barn! We also had twelve acres of land, but that didn't seem so important then.

We have never regretted that bid. I think a lot of people get too caught up in trying to screw down the vendor to the lowest they'll accept and then risk losing the property. When we discussed the bid our biggest fear was losing it, not overbidding, and I still believe that was the best way to go ... and that's in part due to the experiences of someone I knew at work. John was also looking for property outside London, and most especially a barn conversion, or renovation, and had been looking for around eighteen months, so was in a very similar position to us.

The day our offer was agreed I mentioned it in passing, and he told me that his offer for a barn in East Sussex had just fallen through. After a bit more discussion we established that it was the exact same barn I'd seen a year earlier, which had had all those people viewing it. What seemed to have happened is that John got the price down to a very keen level, but then the ill-will in the negotiation had caused the process to grind to a halt, and then eventually fail – and he'd lost the barn.

Unfortunately the bidder for the other lot wasn't quite as organized as we were, and his bid in fact fell through. This affected us because the vendor wanted to sell everything in one go, and so we were all slowed down; during one follow-up visit we shared some of our frustration with him, and expressed our eagerness to get moving. The vendor was a farmer, Ben; he had lived on the land for sixty-one years, and was of the old school in the sense that once we'd made the agreement, it was settled. He spoke to his lawyers the day after we'd mentioned how slowly things were going, and from then on everything started to speed along.

Ben was clearly delighted to be retiring from the back-breaking and often heart-rending work of being a full-time farmer; his wife, however, was very upset at the prospect of having to leave the hearth and home she'd loved for over half a century. Nevertheless I think the extra profit they made from having obtained planning permission on the barn, a process which had taken nine years, meant they were going to live very well in their retirement, and Ben had also retained a couple of fields so he could still keep a few cows. Ben had clearly been running the farm at minimal maintenance for the last few years, which made sense given he was close to retirement, and this left us with minimal fencing – bits and pieces tied together with twine, and all sorts of temporary solutions. Over the years we often used the same sort of Heath Robinson approach, and we always thought of Ben when we did, and hoped he'd approve!

Converting the Cowshed

When the exchange finally happened it was almost a damp squib, having taken so long – nearly four months. We completed a month after exchange, at the end of October, and immediately went to work doing our temporary conversion of the cowshed. We had a basic plan of attack, which was to clear out the cowshed, deploy permanent drainage, and build a temporary structure within the cowshed. We were lucky in that bringing in electricity would be easy, as the line passed directly over the barn and taking a feed from it was surprisingly cheap.

Clearing the cowshed involved knocking down the internal walls, and this meant several weekends of hard graft with a sledgehammer as these were blockwork, filled with concrete and then covered in concrete render, and they really didn't want to move. The underlying floor in the cowshed was concrete, sloping to a drainage channel in the centre, which meant it was likely to be saturated with cow urine and therefore not viable in the long term. However, for our short-term plan it made a good base for the supporting timbers of the floor. Short term was not particularly well defined at this point, but I think if anyone had asked we'd have said around twelve to eighteen months...

We were doing all this work during the winter, working at weekends when we could, and also taking time off to head down to the barn. It was often bitterly cold and our heating was rudimentary – namely jumpers, gloves and woolly hats! The facilities were fairly basic at first: we did have electricity, but no drainage until we put it in! Lunches usually consisted of an old-style Brevel ham and cheese toasty – truly delicious, though we occasionally treated ourselves to a trip to our local pub. The food there is excellent, but it often meant that lunch took two to three hours; still, it fortified us and enabled us to keep going.

One of the worst jobs during this stage was associated with putting in the drainage. We were fortunate that there was mains drainage available, we just needed to get our pipes about sixty feet out into the field, and then a further ninety feet along the edge of the buildings to where we needed it. Much of this was easy digger work, which Gordon loved (possibly related to the covered cab and seat), but about sixty feet of it involved breaking up some concrete which had been laid between the cowshed and the large steel barn (which needed to be taken down at some point). Initially we thought this concrete was fairly thin and would break up easily using a normal electric-powered breaker. Sadly, however, it was three to four inches thick and really tough, which makes sense as it had been the main trackway used by the cows to get out of the barn and into the field, and also I suspect an area where the farmer could corral the cows for inspection or loading into a lorry.

I don't remember how many days I spent in the cold, often rain, with my hands juddering with the impact of the breaker as I slowly created a two-foot wide channel through the concrete. I suspect it was only three, but it felt like a lifetime; sometimes I was so tired I could barely lift the breaker out of the hole I'd just made. It helped me develop real respect for all those workmen you see on roadsides; the reason there needs to be several of them is so they can take turns and keep going, as well as supervision and tea duties. The latter were ably performed in our work gang by Sue, who also pulled a full shift at moving debris, and carting sand and shingle around. Her relentless stamina was inspiring, and occasionally depressing, as it meant I had to get up again to go back to work to try and catch up with her!

All this hammering and digging allowed us to put the drainage in, a very exciting thing indeed, and we started to build up the floors using bricks and beams to raise it up and then chipboard to fix it. We then put down insulation and covered it with chipboard – and that was it! No fancy terracotta tiles for us, no floorboards or carpet, just plain old chipboard until we moved into the permanent converted parts, which seemed a very long way off at this point. All this activity took place in the period up to Christmas, and just before the holidays started we had put in about a third of the floor, some twenty feet or so, which was planned to take a shower room and a separate toilet, and then the start of our bedroom.

Gordon decided that it was time for some proper toiletry facilities so he built a stud frame for the separate toilet, attached the door and installed the toilet – and so our first room was complete! As we had all the facilities we decided to host a New Year's Eve party, warning everyone that it would be cold and that they needed to bring sleeping bags. I think the cold encouraged everyone to drink more to try and stay warm as it turned into one of the most drunken and raucous New Years I can remember (and I can barely remember it!) – but it was a great way to christen the house. We moved in properly less than a month later.

Another lesson of this period was that no matter how carefully you placed building materials when they arrived, you always seemed to need to move them – again, and again, and again! Moving insulation is no great problem, bricks are OK if tiring, but I really, *really* hate moving plasterboard: it's heavy and awkward, and if you drop it or catch it then it shows it immediately, and it's not fixable. After some practical experimentation we developed a technique that required two people to grasp it tightly on its upper edge with both hands, holding it a foot above the ground and shuffling along; any stops involved lowering the plasterboard on to one's foot to take the strain and to make sure it didn't knock into the ground. This was one of a host of techniques that we were all learning on the job as we slowly got to grips with the tasks in hand.

The internal walls were all unfinished plasterboard, and when we actually moved in, at the end of January, we had only three completed (of a fashion) rooms: our bedroom, the shower room and the extra toilet. If we looked at the ceiling we could see the roof tiles in some places and the sky in others, though most was covered by thick black plastic and insulation – not that it seemed to do much good. But we were in! Weirdly, the floor area of our temporary home was almost exactly the same as our old flat in Islington (which we sold shortly afterwards), so we were able to fit in the majority of our possessions. Or as a friend kindly observed, we'd moved our basement flat into a barn and paid a huge amount of money for the privilege!

Moving In

The night we moved in was one of the coldest I ever experienced, not outside, but inside. We spent the day moving boxes in, building the bed, and nailing up some last-minute insulation. Having no cooking facilities as yet we went to the local Thai restaurant and ordered lots of takeaway, and then took it home for our first meal in our new house, a veritable feast with the heat of the red curry helping to keep us warm.

Unfortunately our fish weren't quite so lucky. One of our tank heaters seemed to have stopped working – we'd bought a second tank a few months before, partly for growth and partly as an isolation tank for new fish, and all the fish in that tank had just stopped moving. They weren't floating to the top or sinking to the bottom, they were just stuck at whatever place they'd been swimming, as if they were frozen in place – which to an extent they were! Once we noticed their predicament (after our dinner, and with temperatures plummeting) we quickly fished them out and placed them in the other tank, where the water was much warmer, though at the bottom of the accepted range. Amazingly it was like bringing them back to life, and they immediately started swimming around as if nothing had happened. All in all we lost four fish as part of the move, which while sad, was certainly not as bad as it could have been; so we still had seven, the four harlequins and three recently purchased angel fish, whom I had named Symphony, Harmony and Rhapsody after the characters in *Captain Scarlet*.

After all this excitement we sat on our sofa, which looked tiny in the still open area of what was planned to be our temporary living room, dining room and kitchen. We opened a bottle of wine and then just slumped in exhaustion. I think both of us were questioning our mutual sanity, and as the temperature continued to drop there was also some fear as to our ongoing health! Cold was a particularly sensitive area for us, as we both come from families where the parents had always kept it colder than comfort would seem to dictate. 'Put another jumper on then', was what I was told if I complained about the cold, and Alex heard much the same – I think it's a reflection of the generation born during the war and the austerity of the fifties. . . (or maybe they really did have ice running through their veins and couldn't feel it). I've always felt that if it's cold enough for icicles to form inside the house, even if they are against a window, then it's too cold! This being our mutual background, we had enjoyed heating our flat to a minimum of 20°C at all times. By comparison the temperature that first evening was hovering around 4 to 5°C, and it was clear to us both that this was likely to be our evening temperature for some time to come!

Within a week of moving in we had Sky and broadband installed: we might be in the country but we wanted to have the benefits of technology, and early! But we did not have heating, and this was a really bad omission. The north wall was just a single brick in thickness, and when the north wind blew it was as if the wall wasn't there at all and we were standing outside stark naked and covered in freezing water. As a moving-in present Gordon and Sue paid for a month's use of the space heater we had been using for the final weeks to keep ourselves going, and this was a real godsend. I'd wake up in the morning with my whole body boiling hot but my face freezing cold, so I would then brace myself and run the length of the cowshed (about sixty feet) to switch on the space heater, then run back and dive back into bed before I was completely frozen. After about five minutes of space heating the cowshed was bearable and we could get up to shower, although this was still a bitter experience.

Even after we completed the walls and bought storage heaters the temperature would regularly be as low as 5°C when we woke up in the morning throughout the time we lived in the cowshed.

Planning the Barn

Now that we were in the barn we needed to properly plan the rest of it. We asked a surveyor whom Gordon and Sue had used when they built their house extension to come and take a look around. Some might say that probably we should have done this before we bought the place, or at worst before we started on our temporary works, but we just hadn't had the time, and I suspect we also didn't want anyone to pour cold water on our dreams. And certainly the experience of the surveyor's visit was a little depressing. First he told us that we were on a journey of a thousand decisions (this wasn't depressing), and that our budget was completely wrong (in fact he laughed at it, a lot); he also informed us that we would not be able to recover any bricks from the old dairy which we had to knock down as part of the planning permission (one of my money-saving hopes).

After this we retired to the local pub to talk about it. In any project there is a triangle, the three sides representing quality, time and cost. It is only possible to get two of them close to perfect, and we chose to go with just one – quality. We decided that even if it took longer than expected, say five years, and cost more than we expected, say three or four times as much, we wanted the quality to shine through always. At the time it really did sound sensible!

When we'd bought the place almost everyone had said to us, 'Oh yes, just like *Grand Designs*', although I have to admit that we didn't really know what they were talking about as we'd never watched that particular television programme! We did once we were in, however, and for a while it became one of our favourites and a source for a few of our ideas. Several people asked us why we didn't go on the show, and to be honest the main reason was that I never wanted anyone coming along and poking around our house and judging all our decisions. In addition, and the reason I usually give, the show tends to want projects of a year or so, and we definitely weren't one of those! One thing which amazes me about a lot of those property shows, especially the ones involving a search for a property for a discerning (read 'picky') couple, is the number of people who walk into wonderful buildings and just fail to see the beauty and possibilities within the structures.

Alex and I tend to hold strong opinions and argue about them fiercely, and usually we agree, although when we don't we can get a bit voluble and vociferous. In the course of buying the barn and doing the initial conversion we had come to a very clear view of what we wanted to do with it, both in terms of look and floor plan. While we would tune this as we went along, and define the detail to a greater extent, we kept to this key vision throughout. Early on, however, we needed to get some more detailed plans together, so we resolved to speak to some architects and surveyors. We really liked one architect, but it was clear that he wanted to design the whole thing (as might be expected). This just wasn't what we had in mind, so in the end we went with a surveyor who seemed to understand our intentions. He drew up some plans which we only changed a little, and then we were away.

Our Converted Way of Life

Over the following years our ongoing barn conversion became a sort of background hubbub to our lives. We'd often have to do some additional planning, or take a week off to finish a roof, or some other peak activity, or perhaps take a day off to go and 'source' products. Sourcing was actually one of the more enjoyable activities as it would involve visiting showrooms, builders' yards or DIY shops to view potential materials or solutions for the house. We also needed to keep an eye on the costs, and manage the various contractors we brought in over the years to do specific jobs (electrician being one of the key jobs, especially after the law changed to require a qualified electrician either to perform all works, or at a minimum, review and certify everything). However, despite all this the barn never grew into a monstrous octopus dominating and controlling our lives and trying to suck all the joy out of it: instead it became almost part of who we were, and a generally fun part at that. I think this was because we were building something materially solid and lasting (which as a lawyer and an IT person was in stark contrast to our jobs), and also because we were constantly learning new things.

Quite a few of our friends worried that we'd be constantly dragging them to the barn and then forcing them to work on whatever dreadful tasks they imagined were required in the project, and I suspect that's why they didn't visit us – as well as the fear of the countryside and the huge distance from London, of course... One friend who helped us greatly was Adam: he was with us when we were putting up insulation, he helped when we were moving plasterboard around, and he was a huge help when we pulled down the dairy during our first Easter.

One of the parts of the planning required us to demolish the old dairy which sat in the crook of the L shape of the barn. It was a brick building with a white tile roof, and it presented a real destructive challenge – and entertainment! Adam was particularly involved in the destruction of the side walls, where we would all take turns to smash a sledgehammer into the wall until it finally cracked and fell over. I like to think he enjoyed the exercise! Another friend came to visit during the same period, and I couldn't stop him from smashing down the dairy and nearly had to restrain him as he'd become so excited! However in general we didn't ask people over to help, just to visit and be entertained – hopefully!

We had moved into the barn in Hampshire while we were still both working in London and commuting. My commute was around two hours each way (and much more on a bad day), but coming home, especially in the summer, was like going on holiday, every single day. Still, it took us a little while to get started ... and our first morning was not the best. We decided to get the 07:44 train, as we figured it would get us to work for 09:00 without any problems. The alarm went off at 06:50 and we started our morning ablutions – jumping from warm bed into freezing air, then into hot shower and back out to freezing air and so on – and all seemed to be going well. However, we somehow slowed down and didn't manage to leave the house until 07:30 or so – though still this seemed OK.

But when we got to the station we found we had to pay for the parking, and rushed around getting out coins and buying a ticket before running towards the train ... only to see it pull out! This precipitated a rather loud row in the middle of the ticket office where we both allocated blame, perhaps unfairly, until we calmed down long enough for me to buy a cup of tea and then get on the next train, slated to be the 08:14. This got me to work just before 10:15... The next day we managed to get the 07:44, but this got me to work for 09:40, so our target train became the 06:44...

Waking up at 06:00 to get ready in time to catch the train was a shock – but stepping out to see the fields and the early morning haze was just magical, and more than made up for the lack of sleep. All in all we'd been amazingly lucky to find the place, and we were enjoying living there. And that's where we were when we started our menagerie...

PART 1: THE EARLY YEARS

BOB THE CAT

One of the great things about living in the country is the wildlife. Arriving home after a hard day in the Big Smoke and walking – when on the odd occasion Alex is away and has the car – up the drive on a summer's eve to be greeted by deer, rabbits and the occasional owl or pheasant is quite special. However, there is one place I am unhappy to see the beasts of the field and fowls of the air, and that's inside the house. Most especially I don't like mice or rats, and for a reasonable period of time we had both. Some people have confidently stated to me that mice and rats don't mix, and this may be true when planning a dinner party, but it certainly wasn't true in our house. Of course it took a bit of effort to persuade Alex of both the problem and its eventual solutions, in part because Alex likes them – well, she likes mice, anyway – and she didn't see why they shouldn't have a decent home. I, of course, agreed, I just didn't see why it had to be our home!

A Rat Problem

I first spotted the mice when one casually wandered along the wall plate above the television while I was watching it. Then there was some evidence of food being eaten, holes in boxes and the like. I was told not to fuss, it being the country and one just had to get used to it, and to be fair it didn't bother me much, a few plastic tubs and extra tidiness around cleaning up after dinner (something Alex will claim rarely happened!) did not seem much of a burden to me (especially as it was Alex doing it!). But then I saw the rat. I know it was a rat because it was about five times the size of a mouse and had a rat's tail, which seems all the evidence anyone could ask for.

Now rats I am really not keen on, partly as carriers of the Black Death and other nasties, and partly because they can attack humans (this may not be entirely true, but I'm sure they're nasty, and that scene in *Indiana Jones* was really quite scary) and cause quite a lot of harm. George Orwell's *1984* and Room 101 really didn't help either, and I'm sure most people would break when confronted with a cage full of snarling sharp-toothed rodents. When I told Alex about the rat she laughed at me, and assured me it was almost certainly just a mouse and that I should stop being a BGB ('Big Girl's

Blouse') – I did try and explain the creature's definite ratliness, but to no avail.

For a couple of weeks after that the crafty little critter would wait for Alex to be out of the room before ducking out and waving at me, or at least scampering nimbly over the kitchen work-surfaces. (One key thing from this whole episode was that it persuaded me to be extra clean when preparing food, not something I had been terrible about, but laziness had occasionally crept in...) I mentioned that a cat might be an answer to our problems. In no uncertain terms, this suggestion was rejected on the grounds that a) I was the one getting all wibbly wobbly about a little mouse, and b) cats are evil, stroppy monsters, good for nothing other than plotting to take over the world (we had recently watched the film *Cats and Dogs* which is amusing, if a little prejudicial). The comedian Jack Dee's sketch on the difference between cats and dogs, which was decidedly biased against cats, didn't help the situation much either! Yes, Alex is a dog person.

Personally I am an animal person, and love both dogs and cats (and goats, cows, sheep and horses, but not rabbits, which I consider nasty, or rats, as already observed). But I have noted that people often tribalize (in this context meaning 'to form tribes' – with the author's apologies for this verbalization) about relatively trivial things, and cats and dogs are one of the topics which most commonly bring about this behaviour (the other most common topic being football). Alex was particularly bad about being a 'dogist' – some might even say she was rather dogmatic... – especially as she generally showed little remorse that her few interactions with cats appear to have involved running them over (or nearly). Still, I thought some perseverance might bring her round. I mentioned that a cat's purr will bring about feelings of happiness and contentment, that they are soft and strokable, and that they require little in the way of maintenance. I tried talking about how Siamese cats were talkative and friendly, only to be told coldly that they were the epitome of evil (Disney and his film production of *The Lady and the Tramp* have a lot to answer for!).

Alas, for weeks she would not be persuaded, and kept reminding me of the saying 'Dogs have owners, cats have staff' – though this I think misses the point somewhat, even if it is true. We went so far as to buy some of those subsonic gadgets, which emit a sound at a frequency only rodents can hear and which are supposed to disturb them and scare them away; these worked in the sense of keeping some of the surfaces free, but as they don't go through any surfaces they didn't stop the little beasties getting into the drawers, for example. But alas for the rat, one day it decided to get into the cupboard and jump out at Alex when she opened it looking for some breakfast cereal. This was a truly monumental miscalculation on the part of the rat, and I can only attribute it to overconfidence and some form of adrenalin addiction. I was duly informed we had a rat and something needed to be done! Knowing the wisest way to respond, I acted surprised at the notion of a rat problem, and agreed to start thinking about what measures might be taken. . .

Taking Measures

My new and entirely unprecedented suggestion was that we should get a cat – and not any old cat, oh no, we should get a Farm Cat (I really wanted a Siamese, but the rumours of evil, and my concern they might be more into talking than ratting, persuaded me to avoid the topic). This semi-mythical feline was popularly held to have the power to hunt down vermin by the bucketload (a traditional measure of rats), and also to multiply like a rabbit (although I chose not to share the details of this second mythical superpower with Alex).

After some further discussion and cogitation my proposal was approved, and I set out on my quest to secure our new weapon against our unwanted house guests, using the power of the internet. Entering 'farm cat' into a number of search engines quickly directed me to an RSPCA centre in Kent which had a 'semi-feral cat, about two years old, would make excellent farm cat'. I sent an email to the address noted (the fact that it was an email added to the allure of this particular ad – though I do often wonder why, after using the internet to start a process, people then insist on that old-fashioned medium the telephone to complete it) then phoned up to arrange to pick up 'Becky' from a relieved-sounding woman – perhaps alarm bells should have rung.

Alex drove us to the centre, which was near the mouth of the Thames, about two hours away from us (the geographical disconnection of the internet providing ample opportunity to see the country, albeit from a motorway, though I guess we could have just driven up and down the M3 for two hours to get the same effect), with a brief stop to pick up a cat carrier in Guildford. This stop was not exactly planned, in fact I think we were in a panic as we'd not only failed to consider the requirement for such a thing, but we were running late and already on the road; nevertheless a quick search of the internet (using my Blackberry) found us a pet shop in Guildford. Sitting just outside was a perfect cat-carrying contraption, which I purchased as quickly as the nice lady behind the counter would let me! There were no more stops, and we proceeded happily to deepest darkest Kent.

When we arrived at the address, we were a little confused as we appeared to be in the centre of a housing estate, but we went up to the nice semi and knocked on the door, assuming they'd tell us we were in the wrong county or something equally upsetting; but they were in fact happy to see us, and directed us to a large shed at the back of the garden. The shed opened into a relatively large, well-kept room with a wall of cages on one side. In one cage there was a cat and her recently born kittens, and the nice woman who ran the place was hand-feeding another kitten which had lost its mother.

When we said we were there to pick up Becky she looked very happy – almost too happy, in fact; and when she pulled on two large, kevlar-lined gloves those alarm bells really started to ring. She tentatively opened one of the cages and pulled out a spitting, snarling ball of fury which she very carefully – in the sense of keeping it as far away from her as possible – put into our carrier. She was so relieved that she almost forgot to get us to sign the paperwork and make an appropriate donation. We left in a state of mild apprehension, particularly as her parting shot was something to the effect that she'd never been so pleased to get rid of a cat...

Becky

While we were filling out the forms the woman had given us some history on Becky. She was about two years old (though I'm not sure how this was determined) and had been living on an industrial estate for a while before a man had picked her up and taken her home. She didn't give details, but Becky clearly hadn't fitted in well in the house, which already had a few cats in it, and had been brought to the RSPCA. As we didn't have a cat she thought it wouldn't be a problem, especially as we were planning for her to live semiferally anyway, in the sense that she would be able to roam around the fields and sheds. In passing she told us about a similarly excitable cat she'd had a few years before, which she'd passed on to a new male owner. Apparently he'd phoned a few months later to tell her that the cat was now sleeping on his chest, which was an amazing transformation from a cat that would claw anyone who got too close. We smiled in appreciation at the story, and wondered if she was just trying to soften the blow?

On the way back to her new home Becky seemed to calm down a little, although it still wasn't wise to put one's finger too close, as she was not averse to lashing out. She was a lovely cat though, all black with little white socks, and when she sat still she was 'just so'. Before we even got home we decided that 'Becky' was really not an appropriate name for her, so after some debate we arrived at calling her Bob, and Kate for short. We had read up a lot about how to help cats get settled, and they all agreed that the new member of the family should not be allowed outside the house, and preferably not out of a single room, for at least a few weeks. However, at that time we didn't really have complete internal walls, and any real climbing cat would be able to get out of any room; in addition there were several ways of getting under the floor, not to mention some larger than is perhaps normal holes in the external walls through which getting any cat to return might prove tricky – so we had purchased a couple of puppy cages and put them together to make her an area in our room. This, combined with a few toys and a bed, was to be her settling-down quarters.

Bob Settles In

Bob wasn't entirely keen on the cage, as she showed by making quite a bit of noise during the first couple of nights, but after a while she seemed to decide that we weren't monsters, that the food was OK, and that she could relax a bit, though she totally ignored the scratching post we bought her. This seemed to be a bit of a theme with her, in that she would never scratch any form of scratching toy, and after two more failures we stopped buying them. I let her out of the cage a few times, and she sniffed round the place, climbed on me a bit, and generally seemed to enjoy herself – and getting her back in wasn't the trial I had expected.

After a while we let her out more and more, though always keeping the doors closed, and she became more and more friendly, until she got into the habit of sleeping on me, just like the man in the story. I nearly phoned up the woman in Kent, but chose not to as I decided she might think we were either joking or being mean (and it would involve phoning, which as I may have mentioned, I hate doing).

She was quite good with us, but every now and again she'd snap and try to claw us. It wasn't immediately obvious why, but one of the triggers was stroking her, or indeed touching her in any way, about half way along her back. She was very sensitive there, and it made us think that probably she'd previously been hurt on her back. It wasn't the only cause but it was the most obvious. We had a few words on a couple of occasions and eventually she seemed to understand that scratching and clawing was out, helped I think by us learning to play with her. Alex doesn't recall this part and thinks Bob just became more settled when she realized that she was safe, but I know my little chats with her helped! Her favourite game was for us to tap one of her paws and she'd try and catch the finger tapping it, then tapping another one. If she caught the finger it would hurt as her claws were always out, but after a while when we'd play this game, she wouldn't always use her claws, at least with me. But she still keeps her claws out for anyone else! She also eventually became used to a full stroke along her back, though she always arches her back to the stroke, which is a gentle way of getting her to stand up if she's lying on my chest and I need to get up.

After a few weeks we allowed her a free run of the house and she seemed to settle down well. Each night she'd sleep on us, usually me, but as I am rather a restless sleeper she would often end up on Alex by the morning. As she settled down on my back she'd purr contentedly, and it was a source of great pleasure to both of us to see her calming down so well. We even let her out into the main barn where she showed that my fear she might just run away was completely unfounded, so I resolved to get her a cat flap.

For some reason there is only one real supplier of cat flaps on the market, and their flaps are clunky mechanical things with different forms of opening mechanism in that the cat can carry a magnet, or an electronic tag which should open the flap to them. Unfortunately the placement of the detector requires that the cat approach the flap in just the right way, and Bob never seemed to want to do that – so she'd butt her head against the cat flap and give up. Eventually I just left the cat flap unlocked so that she could move in and out comfortably – and in theory so could anything else which might be interested in exploring our house!

The Continuing Rat Problem

Of course the reason we'd invited Bob to join us was to help reduce our rodent population. I had also in parallel bought some poison which was supposed to kill the rats by filling up their stomachs (rats, I was reliably told, cannot throw up, so if they eat something they cannot digest then eventually they will be so full they die, almost like Mr Creosote perhaps), but wouldn't hurt cats, owls or whatever. Sadly this poison was not working, mostly I suspect because I hadn't placed it in the right way, but also possibly because they had to eat a fairly large amount of it, and they seemed to prefer our food to the poison! Therefore we were left with just Bob to deal with our pillaging pests.

Sadly it turned out that Bob was not a ratter, and that's an understatement! One of the main reasons is probably that she is actually rather a small cat, not really much bigger than a large rat herself. Once when the rat had come out to play I placed her on the work surface. They both looked at each other, seemed to come to an agreement, and it wandered off while she jumped down and headed to look for food. In our first year the only creature I saw her catch was a baby rabbit which was nearly half her size. It was still alive as she tried to get it through the cat flap and I managed to rescue it and release it back in the hedgerow, much to Bob's chagrin.

She wasn't much of a mouser, either. In the time we've had her she's had one three-week period, when we'd had her about three years, when she caught mice every night, which she'd bring into the bedroom and play with until I woke up and took the bits away (yes, it was horrible, and even more so at 02:00). She once even brought a live mouse back to the bedroom and then let it go, to spend the next thirty minutes having a huge amount of fun chasing it around the place, which at 03:00 was really not much fun for me, made even more annoying by the fact that Alex would always sleep straight through these episodes. Still, after three weeks she seemed to become bored, and never caught a mouse again, much to my disappointment (from a mouser perspective) and relief (from a sleep perspective).

Our neighbours had similar problems and had bought some proper poison which the rats had taken a liking to, and for several months we were not bothered by any more little furry visitors. However it never lasts, and eventually they were back again, stealing food merrily. In the end we called out a pest removal service, who laid traps all around the property, and these seemed to do the job! Fortunately by this time Alex had become attached to Bob, or BobKate, or Bobula, or Bobbob depending on one's mood, and would not think of letting her go: and so we had our first pet, and a lovely one she is too!

A Part of the Family

The other step our neighbours took around this time was to provide a home to some cats as well. As is typical of all felines, they took no notice of human defined boundaries, and would often wander across our land. This caused us trouble as Bob clearly wasn't very happy with them, and at one point seemed to have a big fight with one of them. She started peeing in random places, including once on the bed, but often on my T-shirts or trousers (and on a couple of occasions I only discovered this as I put the trousers on...).

We spoke to the vet who said it was a sign of stress, and that potentially her stress might be because one of the cats had come into the house, so he recommended we close up the door to the outside and see how she recovered. After around two weeks she seemed to calm down a lot, and we tentatively reopened the cat door. Clearly she'd learned her lesson because she never had a problem again – I suspect she made sure she went out carefully and avoided the other cats. We tried to help by chasing off the other cats whenever we saw them – but as with most cat-related activities, I suspect they were just laughing at us!

As well as not liking other cats, which we saw when we picked her up, and then saw again when we first took her to a cattery – she was quite definitely not happy to see the other felines strolling around in their cages – Bob has only a few other foibles, two of which involve food and drink. For a cat who allegedly was running wild for a year or two she is enormously fussy about her food. In general she will only eat when the food is fresh out of the packet. She will sometimes eat a part of it and then finish it a bit later, but if she doesn't actually see us when we empty the food into her plate she just won't eat it! It's a wonder she survived in the wild...

Her other habit is perhaps more understandable, in that she refuses to drink out of a bowl or a saucer. We'd put milk down for her and she might sniff at it, but she'd rarely touch it, and it was the same with water in a bowl, until we eventually gave up on it. The vet told us she'd be getting enough water from her food (provided we kept to the wet food), and would make it obvious if she needed more, so we left it at that. Then one day Alex left the tap in the bath-room running (having gone through an obsessive phase about water scarcity, I very rarely, if ever, leave any taps running so I know it must have been her!), when Bob leapt up and started lapping at the water flowing out. Ever since she'll go through periods of jumping up into the bathroom basin and looking at us imperiously, occasionally supplementing the look with a curt miaow until we run the tap for her. We surmised that she had discovered dripping pipes were a good source of water when she was feral, and perhaps that pud-dles tasted nasty? Either way we now knew her preferences and were trained to follow her instructions!

A couple of years after Bob first joined us she was at the vet having her stand-ard check-up when the vet noticed that her teeth seemed to be covered in plaque. We agreed this should be remedied, and organized another visit to have her teeth properly cleaned, for which she was sedated (otherwise I could see her shredding any digits foolishly placed in her mouth!). While they were cleaning her teeth they discovered that several were rotten, and had to pull them out. This, they told us, probably meant that she was considerably older than we'd thought, perhaps as old as five or six when we got her. They also recommended we give her more dry food to help control the plaque, though I hoped she had no other rotten teeth otherwise we'd be causing her extra dis-comfort. She, however, seemed oblivious to the removal of her teeth, and car-ried on regardless, including deigning to eat some of the hard food we gave her.

Bob's routine was now fairly standard: she'd wake up in the morning, stroll along for breakfast, only eating half in the first sitting, then find a place for a quick catnap, usually on a warm spot, or an area of softness. If we leave a T-shirt on the floor, or a towel or mat, then she'll always choose to sleep on that, which then ends up becoming covered in black hairs. After her catnap she'll wander back to the bedroom and jump on our bed, there to catch forty winks or so, and again if there is a T-shirt or some other material, will choose to lie on that. A bit later she might wander into the bathroom and lie down again to get in some serious zees, only waking up a few hours later to reposition herself for another visit to the land of nod. In the afternoon she likes to have a quick kip before settling herself in for a quiet doze, either back on the bed or on a soft chair. She'll stir herself for dinner, and then snooze gently until bedtime… and after such an exhausting day she'll often then sleep on me for a few hours!

Occasionally there is more to the day, in that she does like to sit on the windowsill when the window is open and watch the world go by, and she sometimes goes for a wander outside. Sometimes she wants the day to start early, usually if I want to lie in and she's hungry, and wakes me up either by eating my hair or by rubbing her nose against mine – which does wake me up quite quickly.

I think it's fair to say that Bob has found an easy life!

BILL AND TED: OUR VENTURE INTO GOATKEEPING

The autumn of 2005 was a good one, with one of the best things, for me, being a week with the lads drinking wine on the banks of Lago Maggiore in Piedmont. There is only one event of that trip which I wish to relate, and that involved a phone call from Alex (who was at home as this was a 'lads only' trip – alas the last one to date...). She called to ask if I was happy for us to get some goats. *Goats!*

I must apologise to goats and goat lovers in advance, but I'd not heard many positive things about them. There is even a website dedicated to exposing how evil goats truly are, mostly by comparing them to cephalopods (squids) as they are the only creatures with the same evil eyes: I recommend it – if you Google for goats + cephalopods + evil I think it's the first hit – it's very amusing, and will prepare you for the true nature of any goats you might meet! I had always felt it unfair that at the final call the sheep would be separated from the goats, and then obviously be given preferential treatment – but then goats are evil, are they not? You only need look at their eyes.

By reputation they are also escape merchants, and they eat everything, especially whatever it is you'd rather they didn't. They are smelly, awkward, and even capricious (derived from the Greek for goat – *capra*). So I said yes, for two reasons: firstly because I had at that point already enjoyed a couple of lovely glasses of Barolo, and secondly because it was a rescue mission – could I really let two wonderful goats (as described) go to curry?

A family living in my in-laws' village had two pet Angora goats penned up in a section of their large garden. Unfortunately the two rascals quite often broke out of their enclosure and either attacked the young orchard (bad) or ran into the road (very bad), and as a result the man of the house had put forward that to avoid any tragedy involving the goats and a car doing 60mph, a better home must be found. In fact what he actually said was that if they broke out again he'd see to them with his shotgun, but it's much the same thing; whereupon his wife placed an advert in the 'free ads', hoping to find a suitable home for the terrible two. After a couple of responses, from curry houses (these goats were seven years old at the time and so their meat would be tougher than old boots – ponder that next time you enjoy a 'meat' dupiaza) she had begun to despair. She mentioned it to my mother-in-law, who thought we might just be crazy enough to take on the challenge, and that we also had enough land to keep them more than happy (while potentially true, it didn't quite work out that way, but more on that much later).

Thus we rode to the rescue. Our overall fencing situation was rather holey at that point, so the initial plan was to set up some electric fence around our existing sheds and then go from there. The main shed they would be in had a well in it, a very deep one, so we covered it over with some wooden boards and placed some heavy stones on it, and also fenced off the rusted bits of old well machinery at the back of the shed. We then scattered straw on the floor and topped it off with a small fence-mounted water trough.

The Goats Settle In

The jolly day of their arrival dawned, and the two woolly sweaters with legs were delivered, and then led (well, dragged) to their new quarters. It is diffi- cult to explain how we felt about them; they seemed much larger than we expected, and woollier. Handling them involved holding their heads and grabbing their beards (for girls they had very fine long beards!) and then mov- ing them forwards. We didn't grab their horns as these seemed rather delicate and they quickly moved their heads away from our hands in a way that sug- gested they didn't want to be held by their horns.

The amount of effort required was rather high and I was concerned about hurting them, or them hurting us, or some combination thereof; however, we were able to get them moving forwards, if slowly. Much 'maaing' ensued, which was resolved by offering some goat mix – a molassed mixture of oats and grains, almost good enough to be our own breakfast. Even so, it wasn't quite enough, as they seemed ready to eat anything else they spotted on their way, including trees and random plants, so it wasn't entirely a smooth operation – but eventually we got them into their new home.

Their previous owners were clearly upset to see them go (well, apart from one), which gave me a bit more confidence in the venture, as I still hadn't quite reconciled myself to my original assent. But my doubts didn't last because after only a short period I began to enjoy their company.

Our new goats had quite distinct personalities and we dubbed them Willemina and Theodora, or Bill and Ted for short (after that dreadful/ classic – depending on who you are – film. . .). Our gender blindness to names, which started with Bob, was once more in action here, though we used the fig leaf of their full feminine names to avoid too much opprobrium. Bill was clearly the boss, always pushing through first to get to the food; she was also very inquisitive and friendly even when food was not forthcoming. However Ted – or Twisty Ted, so named because one horn was twisted out and damaged – was nervous, and was never happy to be stroked, or even for me to get within a couple of feet; it took some time before she'd even approach me when I had food. I perhaps made her more nervous, because for at least the first month whenever I went out to check them I'd religiously inspect them all over by the light of a torch to ensure they were fine and that everything was in working order. This was in part due to my nervousness at looking after them, and also out of concern for their health, exacerbated later on by tales of fly strike from Matt the shearer and his girlfriend. It's difficult to say what exactly I was looking for, as I really had no idea, but I assumed that if anything looked wildly different or if they looked as if they were in pain or uncomfortable, then that would be enough to warn me of any issues! They, of course, were fine, and we relaxed a bit with them, which helped them relax a bit with us.

The question of what they were for was never really answered, and still isn't to this day. They are reasonable for keeping the grass down, but they rip the grass and don't keep it as level as sheep, though in pasture that doesn't matter too much. They were too old to eat, even as stewed curry, and milking them was out of the question as we would need to get them pregnant, and that was more effort than we were willing to commit to at that point. Which left their fleeces. Angoran goat fleece is used to make mohair, and led me to dream of home-produced jumpers and the like. My mother did at one point suggest to Alex that she get a proper hobby, such as making her own yarn and knitting. Alex, however, will use a stapler on her hems if they come down, so bad is she at sewing, so making knitted jumpers and socks was never going to be a viable option; indeed, I have had to do the sewing repair jobs for the last few years, ever since she learned I could sew! However, as we discovered a year or so later (and having stored one shearing for nearly a year), only cuts from younger goats can be used to make mohair, and the rest isn't really that useful, except perhaps as mulch.

Basic Goat Management

It was fairly late in the year, and Bill and Ted still had their summer growth of fleece which needed to be shorn to give them time to build up a proper fleece for winter. After a slight panic we managed to contact Matt the shearer, who agreed to shear them for us. Matt is a shepherd by trade and had only shorn a couple of goats before, though he had shorn a ridiculous number of sheep. He arrived with his kit and set it all up, and I managed to part coax and part drag Bill over to him. He was much more robust with her than I, and soon had her 'sitting' against his legs while he sheared her. He was surprised by how little fuss she made, as goats are apparently known for struggling and literally screaming when being shorn; and I was surprised by just how quickly he finished her.

That left Ted, who was starting to get better with me, but wasn't quite at the stage where she'd follow me without encouragement. I showed off my 'chasing crazily around after a goat' skills for about five minutes before tiring of the game, catching her, and passing her over to Matt. Ted was less willing than Bill and struggled so much that Matt accidentally broke off her twisted horn. To be fair to Matt it was already half broken, but the sound she made, of what sounded to me like pure pain, was horrendous. He assured us that all we needed to do was spray it, and it would soon heal, but we were dubious, I must admit. He also cut her, causing a wound – he called it a 'scratch' – about an inch long, which he also assured us would heal with minimal trouble after spraying it thoroughly with antibiotic spray. The scratch actually healed astonishingly fast, but the horn was another matter, and bled for nearly a week afterwards. We sprayed it every day until it stopped bleeding, though it didn't seem to bother Ted at all after the first day or so, and then it healed up fine. Thus we had our first brush with damage to one of our animals – but needless to say, not the last.

Matt also showed us how to trim their hooves. The goats had had a number of owners, and at least a few had not trimmed their hooves so they had become horrendously overgrown and misshapen. Alex also thought Bill had an affliction called 'strawberry hoof'. I don't recall if it was, but it didn't matter because the treatment was the same as for a normal overgrown hoof: we needed to try and trim it back to the proper shape. Goats normally live on rocky terrain, which wears down their hooves naturally, but on English pasture there really isn't enough hard material to do the job. Matt tried his best, and also showed us that we could be fairly aggressive in the cutting. We had made one attempt previously, but had been far too tentative.

When one has completed trimming the hooves the normal thing to do is to cover them with a special spray; we use one called Foot Master. It's really just an antiseptic spray, but can help to reduce infection, especially if you've been a little excitable with the cutting and have damaged a blood vessel, which happens more often than one would hope. Foot Master, and most other antiseptic sprays, also contains a coloured dye, often purple, so the area that has been sprayed is clear to see. It is also extremely hard, and perhaps even impossible, to use it without getting it on one's hands, trousers and shirt, and I have turned a couple of perfectly reasonable casual trousers into 'work' trousers by being a bit careless with one of these sprays ('work' trousers being those I really shouldn't wear when I appear in public, with apologies to those who use our local Sainsbury's). In fact clothes are fairly easy to deal with, the real problem is hands, and on a couple of occasions my face, because even if you manage to get to a tap reasonably quickly in order to wash it off – which isn't always practical when you're trimming the hooves of a couple of goats in the middle of a field – it still leaves a residual stain.

We have both gone to work several times with stained hands, and while most people don't comment, perhaps just giving us a funny look, there have been a few who have asked questions. They have usually stopped asking quite quickly when I explain it is to help prevent foot rot! In later years we started using an antibiotic spray with a light blue dye, and then we'd have multi-coloured hands which were even harder to hide. I did start to use a glove, but this wasn't a foolproof solution as the spray would get through any holes, and I'd often have to take the glove off to feel the hooves to check progress and clear them out. After a while I became somewhat complacent about it, and now just frown at anyone who asks impertinent questions or stares at my tinted skin.

After Matt had completed our lesson in hoof trimming he left us with two rather bedraggled-looking goats. When we first got the goats they had full fleeces and from a distance looked like behorned sheep. Once shorn they looked far more like goats, but also much smaller and thinner as the fleece gave them quite a lot of bulk. They not only looked different to us, but they evidently looked, and more importantly smelled, different to each other, too, and this lack of recognition meant they were suddenly with a stranger and had to fight for dominance again, which involves them rearing up and smashing their heads together. This they would do for a week every time they were shorn. Imagine if we humans did the same every time we washed and changed clothes... (actually I do know some people who seem to act like that – perhaps I've found the reason!).

Up to this point we'd been feeding them either by hand, from a saucepan (which had somehow become our feed carrier of choice) or by throwing the food on the ground. We weren't entirely happy with this approach as the feed went all over the place, not all of it was eaten, and it sometimes landed on poo! The solution was to build some goat tables ... we were in the process of knocking down the old dairy at this point and it was yielding some decent slabs of bricks (the cement mortar held them together quite solidly), so Gordon made two piles of bricks and slabs, each with a relatively flat top, which we could use as goat tables. The idea was that we would feed them on these piles, and also that they would climb on them and wear away their hooves a bit more.

You might wonder why we were feeding them at all when we had acres (literally) of grass for them to feed on. Our reasons were twofold: the first was to get the goats used to us and to make them happy to see us, and the second was to give us the opportunity to inspect them fairly close up every day, which would help us spot any unpleasantness such as fly strike (see later), early enough to have a chance to deal with it.

The New Goat Shed

One final event of those first few months with the goats was the building of the new shed. On their very first night the goats had climbed and pushed through the fencing I had so carefully placed across the back of their shed, and settled happily one to each corner. Our fear was that in moving around the shed they might dislodge the covering of the well enough to get stuck, or even fall in. In addition the shed was right next to our boundary with our neighbours and we felt it a bit unfair to subject them to their bleating at such close quarters. So we resolved to build them a new, first-class abode.

Having looked at the prices for animal shelters and blanched, I decided that a garden shed would do them splendidly – so off we went to B&Q, and through some stroke of luck got a double discount on the shed we wanted. Alex and I then spent an afternoon putting it together, using bricks recovered from the barn to build the base to lift the shed off the wet ground so it didn't rot, with a certain amount of swearing and arguing to provide motivation. When we had finished we strewed straw on the floor, and tried to encourage the goats into their new home with some food, and a running commentary on all its new benefits: natural light from a window, modern construction, excellent aspect, fresh straw...

Allow me to digress briefly to comment on straw – Alex and I differ markedly on how much straw to use. In my view she is overly generous, providing piles of it so the goats sink into it, and indeed could hide in it, ready to pounce on some unsuspecting walker. She, on the other hand, views my efforts at straw strewing as truly miserly, and complains that such a bed provides no real comfort to the goats. In fact neither is really of any consequence, as the goats tend to eat most of it anyway!

By this time it was starting to rain, and despite our encouraging them with some lovely goat mix, the goats ran to their shed – the old one, which we had locked. Alex and I felt rather put out by their ingratitude – after all, the new shed had far fewer holes in it, and a window and nice new straw. After further somewhat heated debate, we resolved that they *would* move to their new shed that night. Our initial attempt at persuasion was to leave them out in the rain (they were already drenched, as we were, so it wasn't quite as heartless as it sounds). After listening to them bleat piteously for about twenty minutes, which seemed even longer, we robustly encouraged them, in the sense of dragging, back to their new shed. After a couple of times Bill seemed to get the message, but Ted would not be told, and in the end, after an hour of being drenched, we gave up and let her back into the old shed.

The following day I locked it up again and closed it off with electric wire, and that seemed to get the message through. It wasn't the loving gratitude I'd expected . . . still, better that they were in their comfy new shed than down the well in the old one. This showed that, despite being animals, they had a sense of where their home was, and that once they were settled they saw no reason to move! It was an important episode in changing our views of the goats, and animals in general. We started to understand that they were more than just meat-based machines, more than just soulless automatons only good for cutting grass and the curry pot.

After a few weeks they seemed at least to be tolerating their new home, and while never expressing their belated thanks in so many bleats, I'm sure they meant to.

Providing Water

We'd mastered feeding fairly early, with the goat tables an optional enhancement on the process, and we'd now provided them with proper shelter. Supplying water was much trickier, but still very important as all the authorities agreed that the goats must have access to clean and fresh water at all times. In fact this necessity turned into a nightmare that nearly put us off the whole animal caring business altogether, and when I started writing this I'd actually forgotten the full tedium of it.

When we took on the goats we'd received an additional item, a white water container with bars on the back to allow it to be hung from a fence, designed for use by horses. Our first problem was that we had no fence from which to hang it, so it often sat on the floor where it was easy for the goats to knock over. It was also a difficult shape to carry, and obviously quite heavy when full. This would mean that a successful water delivery job would result in a half-filled tub for the goats resting against the fence, and a mostly drenched Alex or Jason. The goats didn't help, as they thought it might be food and would butt against the tub as we carried it in.

When winter came it was worse, because whatever water had been left by the goats would freeze in the bucket, making it cold and unpleasant to carry, and the ice would slosh out when we refilled it. We improved the situation somewhat with a black rubber water trough, but it was still a wearisome chore. Every morning and every evening for probably six months we trudged out carrying the water, often the pair of us performing the task together, one carrying the water, and the other the food (to distract the little monsters). The worst of it was the almost daily incidents of dropping the water bucket, or spilling it all, or having a goat butt it out of our hands. Really, it was hard and distressing work.

Fortunately the other joys of owning the goats, and their happy way of greeting us (well, Bill's, anyway) overcame our water-related misery and we were intent on solving the issue once and for all! The solution – and let's be honest, a very obvious one – was to install a water trough in their field. This required an available water supply, which was ably provided by Gordon: he took a pipe from the main supply coming into the house, laid it under the floor of the end barn, which we were working on at the time, and connected it to the end of the barn ready to be fed out into the field – he even added a tap so we could wash our hands! It also required the weather to warm up somewhat, as the sub-zero temperatures meant that no water would flow even if we got the pipes out into the field.

Spring finally arrived, and I was able to dig a trench from the end of the house out to where we intended to put the water trough – I intended the water to run for as much of the year as possible, and hoped that burying the pipe would protect it from the cold. We purchased a rather large drinking trough from Scats (a country supplies store), which came with all the bits required including a ballcock and pipe connector, and rather triumphantly I connected up the water! After another hour of redoing all the joints and tightening up everything so that it stopped leaking, I could stand back and watch in relief as the trough slowly filled up. Never again would we have to trudge back and forth with buckets of water, and on that happy day the white horse bucket was consigned to the shed and we could wipe all thought of that monotonous task from our minds!

Experiences with Fencing

Another chore, though one which was generally less painful, was fencing. Our initial stab at fencing was using mains-powered electric, on the advice of a farmer friend. We set up the fencing for Bill and Ted when they arrived, and at least at first they respected it. Our electric fence consisted of three wires strung round a set of metal posts and rods – posts for the corners, and rods to help it keep its shape. The lowest wire was about three inches from the ground, and the next two were a foot and two feet above it. The gate was some wire on a springy catch, though we usually just climbed over the electric fence – gingerly, as a shock delivered around the upper thigh area is not that much fun.

The fence was powered by an 'energizer', which in effect sends a pulse of energy along the fence every second or so (by setting up a voltage of up to 10kV). When electricity is run through muscles, say in the hand, it causes them to contract, which would cause the hand to close and grip ever tighter as the electricity runs through it – which is why when touching electric lines that are expected to be live, it is always best to use the back of one's hand (or avoid doing such a crazy thing altogether). The electric fence avoided this problem as the jolt was only brief and would therefore give one time to withdraw one's hand, mouth or whatever was being used to investigate the fence!

I tested it fairly regularly, at first deliberately with a machine tester, and then more often, and with less thought, by electrocuting myself by climbing over it as described above. According to the tester I had about 7kV in the fence, but I have to say that according to me it was a considerable shock! This gave me some confidence that it would work – and also some incentive to try and avoid getting zapped, although this is tricky if you're trying to carry feed over it.

The fence certainly worked on the neighbour's dog, as one day when she was wandering around near the goats (we didn't have a boundary fence as yet, and she viewed it as part of her yard) she touched the fence. It was a good contact with the wire, and it was raining, which meant she took a full jolt! It caused her to yelp loudly and flee home in a flash, and she never came near the goats again – I don't think she knew what had happened, but she knew it had been near those goats! I think they were rather amused by the incident, but this observation may be just me pandering to their mythical evil nature!

Another of our neighbour's animals also had an incident with the fence – one of their horses. Claire had only recently taken up horse riding, and one of her new horses was a large beast called Tom. He was a rather feisty creature, and on the day in question had managed to pull away from her control. Normally this was just an annoyance, but for some reason he decided to go off on a frolic. He ran round their house, past the barn and then across the fields in a large circle heading back towards home. Unfortunately in his path lay the goat area, and as he was galloping along rather fast, he didn't see the fence until the last minute, whereupon he tried to stop, didn't quite manage it, and hit the fence full on. This gave him a considerable shock across his forelegs, causing him to jump back and run back home, where I think Claire sensibly decided to put him back into the stable with some fodder to calm him down. I wasn't there when it happened, I was just told the story when I enquired about the huge sliding footprints leading to the slightly bent electric fence!

The electric fence also did a good job on our first cockerel, Sonny (of whom more later), who had been hopping happily through the fence every day until one morning he caught his foot on it and took a hit. Unfortunately he was now on the wrong side of the fence and in a panic, but to his credit he rushed back through bravely – and never came near the fence again, preferring to dart through the gate if he could!

However, the real purpose of the fence was to keep the goats in, and to a large extent it worked, especially when I realized they viewed it as a line of convention as opposed to an actual barrier – a sort of border between goat country and the other lands. This they demonstrated to me on many occasions, but two spring to mind. The first was after I had noticed one morning that one of the lower lines on the side away from the house was broken, leaving quite a gap. I was surprised the goats hadn't escaped through it, and made a mental note to fix it the following weekend. Bill, who was always the more adventurous of the two, must have sensed my plan, because the following

Saturday we were woken from our lie-in by rather loud bleating – in fact so loud it was as if it were just outside – and sure enough, there she was in the courtyard outside our bedroom looking around with interest and occasionally munching on something – and it didn't seem to matter if it were a weed or a pot plant. That is, until she found the small trees we'd just been given: apparently they were delicious to goat kind, and by the time we got out there Ted had joined her and they were both contentedly chomping on the trees.

So we dragged them back to goatland – and this was challenging as it involved catching them, and then manhandling them back to their homes, while they persistently tried to get back to the delicious leaves. Since then I have learnt a number of techniques, the majority of which involve bleating and goat mix, which are much better and far less effort than any other method, though alas it took me some while to master them. After we had them back in their area, Bill and Ted looked at me somewhat disappointingly – in the sort of maiden aunt sense. I fed them, fixed the fence line and went back in.

Less than five minutes later there was a bleat and the sight of a goat happily chomping at a rapidly shrinking tree. I rushed out and managed to get them off the trees. Alex was back inside trying to work, and it was a challenge to get her attention while I was still protecting the trees: with two goats and two trees I was outnumbered and had a wide area to defend – and they are certainly crafty: Bill's favourite method was to pretend to ignore the tree and munch the weeds in the courtyard while slowly making her way round to come at the tree behind me. Ted was more direct, waiting for me to be distracted by Bill and then rushing in. Once I had Alex's attention she quickly ran out, enquiring as to why I had let the goats into the courtyard again, and helped me drag them back to goatland, as by this point goat mix had lost its allure for them. Once they were safely behind the fence I looked again at the wire, where it appeared another piece had come apart. I fixed it and returned to the house – looking back at the goats I could see them frowning back at me. Five minutes later they were back.

The problem is that over time, wire fencing degrades: either it would sag to the ground, thereby shorting out and losing much of its potency, or the grass would grow up and touch it, with much the same effect. Indeed we had to check the posts for cobwebs as we had the occasional spider who managed to make a circuit between the metal of the post and the wire, with terminal effect for the poor spider.

Over the course of several hours I was obliged to tidy up the fence, trim all the grass touching it, tighten it up and generally completely fix it. I discovered that the goats would wait until I was out of sight, and then run to where they had last seen a break in order to get out. I also discovered that when their fleeces were thick – and they weren't far from needing shearing again – they didn't really feel the charge unless it was close to full or right on the edge of their nose. But eventually the fence was secure and they seemed either to be tired of all the fun, or unable to get out. I believed it to be the latter because they didn't get out for several weeks, and as I was being far more meticulous in my care of the fence I became confident in its powers to hold them in.

A few weeks later we had a group of friends to visit, and as I confidently showed them our lovely goats they all 'oohed' and 'aahed' in the way city folk do when they see one of their number cut down by some form of rural madness. They mostly didn't want to get too close to them but were happy enough, when coaxed, to hold out a handful of goat mix to Bill, who clearly loved all the attention. We then headed back to the courtyard to start our barbecue, and the city types could breathe a sigh of relief that they'd survived a country experience.

However, Bill had other ideas: it seems she liked being petted, and extra feed of course, so much so that she wanted more – so minutes later she came trotting into the courtyard again, to many cries of wonderment and perhaps a touch of fear, with Ted just a few steps behind her. I decided the Londonites needed toughening up, so persuaded them into helping me to catch the two capricious beasts and drag them back... and catching them again and dragging them back... and again, and again. Each time I fixed a bit more of the fence until eventually, over an hour later, the goats were back in and not getting out, and we were all exhausted. Still, there were some triumphant grins on people's faces, and they all felt a little more 'country' – Man 1: Goats 0! Of course the goats felt the score was more like Goats 10: Man 1, as they had not only had lots of entertainment, but had been fed with loads of extra goat mix.

Goat Likes and Dislikes

It was during the first escaping episode when our London friends were visiting that I tried to deliver some countrified goat lore to the city folk in an effort to show how rural I'd become, specifically in relation to goats and rain. We had discovered that goats hate the rain, and will run back to shelter as soon as it starts to rain (notwithstanding when we moved the sheds, which was, now that we knew them better, a sign of how big a dislocation it was for them). Actually they really only used the shelter when it rained, and tended to ignore it most of the rest of the time. There were many frosty mornings when I'd go out to feed them and they would be lying in the field. They'd get up and run towards me, leaving green patches in the frosted pasture, and if I touched their fleece it would be covered in icy particles; but if I pushed through to their skin I'd be rewarded with real warmth.

They were much the same with snow; it was really only rain they disliked. On this occasion one of my friends was worrying that it was going to rain as we walked back to check that the goats were still in their designated area. Pointing at them, I portentously proclaimed that as they were still out and about (and fortunately still where they were meant to be) it wasn't going to rain, as goats could sense such things. Just as I finished this sharing of a rural secret, the heavens opened. As we ran back to the barn I noticed the goats standing in the field contentedly chomping away at the grass – in fact they only started to make their way towards their shed when we were safely back inside. I swear they did it deliberately, and arranged the rain too!

When the goats were delivered there was mention of them liking ash leaves. I have to admit this completely passed me by at the time, probably because I couldn't have identified an ash tree! The other tree which at that point I couldn't identify reliably was a yew, which had stuck in our minds as a potential source of death to goats. Once we'd dealt with the basic fencing, which didn't actually encompass any trees, I determined that we should know what a yew looked like, and in passing I remembered to take a look at an ash tree. We were in luck: we had no yew trees around our field (and indeed none anywhere on the property), but an abundance of ash trees! I'd sometimes let the goats out of their confined area so they could enjoy the better grass (this could be seen as taunting them, but I really meant to give them a treat), and after discovering we had some ash trees I tried breaking off a bough and feeding the leaves to them. They went crazy over it! Ted rapidly overcame her habitual reserve to get at the leaves, even pushing Bill out of the way in her eagerness. Bill surprisingly took this quite calmly, and just pushed her back in turn to get to the tasty greenery.

From then on it was a special treat for them – and more importantly, a foolproof way of getting them to follow me. Many times after that I could be seen tramping across the field with a branch or two of ash in my hands, and two goats almost gambolling behind me in their happiness at getting to the wonderful treat of ash leaves!

Conflicting Advice

When we had been given the goats I had visited Amazon to gather some reference literature, and in the first purchase round I selected three books on goats. As I was to discover, all animal books quite logically follow much the same format, covering housing, feeding, breeding and what the animal is used for – usually to be eaten – then illnesses and injuries, generally in that order. The only thing that changes is the thickness of each section: for example, in pig books the 'eating' section is often as big as the rest put together, and in one chicken book while the egg section is small, there is a rather large recipe section. One other thing I discovered about animal books was that they tend to contradict each other...

I come from a science background – in the sense that I studied sciences at school and tend to have an empirical view of the world – so I also tend to expect that statements in factual, and indeed instructional books will be based on researched fact, and if they are open for debate this will be clearly stated. Most books on animals are based on the experiences of one or two keepers, leavened with some anecdotes and occasionally provided with some scientific data (to be fair the more commercially minded books have a lot more science in them). They are usually internally consistent, but as keepers' experiences differ they can give rather contrasting advice. One particular example occurred early on with the goats.

One book advised that goats are very clean creatures, and prefer new, clean bedding which *must* be cleared out each week otherwise they will become unhappy, and they may be at more risk from lice and suchlike. The next book advised that goats like to sleep in thick bedding, preferably built up over several weeks, including their faeces and urine, as it provides a comfortable and warm bed for them to lie in, and that naturally its fug helps to provide extra warmth during the winter.

We started off by following the advice of the former, and ended up with the latter, especially as the goats would often try and eat their new bedding as we laid it, but then seemed more than comfortable on a thick pile which might be weeks or even months old. So our approach was to check their bedding regularly and assess how it looked. If there was a reasonable thickness and it wasn't too mucky we'd leave it as it was; but if there was virtually none left, or it was basically just all poo, then we'd replace it. If we had the time, or if it was particularly bad, we'd clear out whatever was in the shed and get down to bare wood and then lay the new straw, otherwise we'd just spread it out over whatever was there. Apart from providing the goats with some additional entertainment as we scraped out the thick layer of poo and straw, I don't think it made the slightest real difference.

The books also made mention of measures to ensure that their hooves would wear down naturally. Actually they all said that in the natural world this would happen, but on pasture we'd just need to trim them. We were not impressed with this defeatist attitude and decided that some innovative thinking was required. We already had the goat tables, and to be fair Bill and Ted did sometimes stand on them, and certainly rested their front hooves on them during feeding – but it couldn't be said that they were wearing them down very much. There was another area which they visited quite often, and that was the gate where we let ourselves into their enclosure, which was also right next to the water trough. Ah ha, we thought, if we put some concrete down in that area, then they will surely wear their hooves a bit more as they greet us, and whenever they go to drink. Thus was born the goat patio!

We moved the fence to allow us space around the normal entrance and mixed up and poured around six mixer loads (a standard builder's measure of concrete) into a wooden framed trench which we'd half filled with bricks and rubble. After two days drying we had a nice, relatively flat but also rough slab of concrete to present to a somewhat curious Bill and Ted. They didn't even notice it, happily trotting over to us, and on to the patio, whenever we brought food. It was with much anticipation that we checked their hooves a month later, expecting that we'd have little to do. Sadly the patio seemed to have made little difference to the actual wear as far as we could tell, and we bent to our standard trimming regime a little humbled by the experience.

Goat-related Problems

We also learned about two goat-related problems: fly strike and bloat. The first we actually had to deal with; the other didn't bother us directly in the first few years, probably because of the preventative measures we took.

Fly Strike

The first potential problem we needed to keep an eye out for was fly strike. When Matt had been shearing the goats he and his girlfriend had told us some chilling tales about fly strike and made us worry rather a lot, and also act to limit it. Fly strike occurs during the summer months and is when blowflies lay their eggs into the skin of the goat, either in a wound or sometimes when they are wet or have a mucky back end. They usually hit during periods of sunshine immediately after rain, though this is not exclusive. The eggs hatch quickly and the maggots start to eat the animal, releasing a signal which encourages more flies to land and lay eggs; if untreated the condition is likely to kill the animal, in a nasty and horribly painful way. If caught early it can be treated and the animal will soon recover, but it is obviously something best avoided altogether.

The impact of fly strike can be mitigated to an extent by spraying the goats with a preventative solution, effectively a sort of liquid fly spray which sticks to their fleece and discourages (and potentially kills) any marauding blowflies. The particular type we obtained was called Crovect, and it has the particular advantage that it can also be used to treat an attack of fly strike and will kill the maggots quickly. Later experience would show us that when Crovect was actually applied to an area of fly strike the maggots would go crazy and literally boil up out of the wound, which whilst utterly disgusting was a good way of cleaning it out. Other protective products do not have this facility, and we might have found ourselves trying to pick out all the maggots, or resorting to soap or something else that was ineffective, and potentially losing a goat because of it.

Crovect is normally used to treat sheep, which are also susceptible to fly strike, and when we first bought some we selected the smallest container available, which would treat a minimum of fifty ewes. (Thus we came upon one of the enduring problems of being a smallholder: there are rarely small containers of anything to do with farming, and when there are, they are wildly expensive!). I thought I had read the instructions clearly, and came away believing the best way to apply Crovect was using a syringe to pour a line of the sticky blue stuff across the back of the goat.

Unfortunately this was completely wrong, as Matt the shearer informed me on his next visit, and in fact Crovect would only protect the fleece it was immediately poured or sprayed on. This was unlike a product called 'Spot on', which we used to get rid of lice, which is administered in a single spot between the shoulder blades and then works its way over the animal from there (which may have been the source of my confusion). Crovect is strong stuff, so it is recommended that you wear gloves to apply it, and that any splashes are washed off immediately. Matt told me once that he could feel the slight burn of Crovect on the fleece of sheep he was shearing even a month after the Crovect was applied – but I guess that's the point of it!

Our friendly farmer friend who had recommended Matt and who would later help us in some of our other endeavours, was kind enough to recommend a Crovect spray gun – and was then even nicer and bought us one! This screwed on to the top of the bottle and had a kind of pistol-grip pump containing a reservoir, which would spray out and refill as part of one motion. When it worked, the liquid would come out in a nice blue spray, and if we could get the goats to stand still properly it would go on in a wonderfully clear and even layer… of course, with us it went on in splotches, requiring us to spray on more than the standard dose.

We took special care to spray their hindquarters, as a mucky back end can attract the flies and give them a suitable place to lay their eggs. The good thing about Crovect is that the blueness fades very quickly, which hid our lack of expertise with a spray gun within only a few days of our applying the spray, while leaving the goats protected for around six to eight weeks – whereupon we'd have to go through the fun again!

Bloat

The second potential problem was bloat, which also tends to hit when it is hot after a period of rain. In these conditions the grass tends to become rich and damp, and when in the rumen of a goat (or a sheep or cow) the grasses start to ferment and produce gas. If it isn't released it can blow up the rumen into a tight ball, and if not treated can stop the animal's digestive process and eventually kill it. The main symptom is that the rumen, which is on the left side of the goat as seen from behind, feels as hard as a football to the touch, and the goat looks distinctly lopsided.

One sunny Sunday afternoon, not long after I'd read about bloat, I noticed that Ted looked a bit odd, and I realized it looked very much as if she'd swallowed a football. Although I was worried that I'd contracted a version of the student doctor's condition of seeing everything as a rare tropical disease, I still decided to check her and make sure she was fine. This being Ted, catching her was not easy – although on this occasion it did seem easier than usual. I felt her side, and sure enough it was rock hard to the touch – I was definitely dealing with a case of bloat!

Treating bloat consists of a few basic steps before anything more drastic is required – the most drastic being punching a hole through the rumen using a needle and the outer container of a ball-point pen to let the gas out. The first step is to walk the animal round to try and loosen things up. Obviously chasing Ted around counted as this step. The next step is to rub their tummy vigorously, especially the rumen, to encourage the gases to escape. The easiest way of doing this is to stand behind them, then to lean over, and using one arm to hold and steady them, deploy the other in a rhythmic massaging motion.

Now picture this: you are walking through a field, and you spot a man bent over a goat (or, say, sheep) rubbing it enthusiastically. I suddenly realized where some of those rumours might come from. Even worse than this was the knowledge that a good outcome to this was for Ted to burp – into my face – and maybe to fart on me as well! Still, ten minutes of this, with the occasional break to walk her round and rest my aching back and arms, and she was starting to burp like a master (and I shan't comment on the other), and soon she was fine. After that first incident there were a couple of times when I thought she might be bloated again, but she seemed to walk it off and never was as bad as the first time. I wonder if I scared her into avoiding a repeat?

Goat Digestion

We also picked up other bits of folklore about goats, the most important concerning their digestion. We were told that in effect if a goat stops eating and digesting and sending things through its stomachs, it will die. It's not immediate, but once the process has stopped it is fairly terminal. This explains their ravenous hunger (or perhaps it has evolved because of it) and is something to keep an eye out for. I think everyone knows that when an animal, or indeed a human, stops eating, then something is wrong, but in a goat (and in most other ruminants) it's very bad. Fortunately we didn't have any issues like this, but we have been told that if we ever see signs of this we should feed the goat either goat's yoghurt or goat's milk (or both). Cow's milk won't do the job, but goat's milk obviously has all the nutrients they require and it can get them restarted if applied in time. For the first month or so we considered keeping some goat's milk on hand at all times, but as we didn't see any sign of the goats losing their appetites (in the least) we stopped worrying. (Ironically, since then we ourselves have converted to goat's milk due to discovering we both have a mild lactose intolerance...)

Confidence Grows

Still, despite all these trials and tribulations, we were starting to feel confident that we could look after the goats. Fencing, trimming, feeding and watering, we'd mastered (or perhaps journeymanned) it all, and we could step back with a feeling of contentment and state with (some) certainty that we were finally and properly country folks, or at least no longer pure city types. However, not being the type to be happy with one victory, we wanted more, and that meant more animals ... and I wanted creatures which would actually provide some benefit, and I love eggs, so it was chickens next!

A CHILDHOOD DREAM: CHICKENS

Ever since I'd been a child I'd wanted to have chickens. This may seem odd, it's not really up there with fireman and astronaut, but it's true (though I suspect if I'd thought about it, being an astronaut with chickens would have been very attractive – I can just see them floating about and clucking in their individual air bubbles...). Actually I wanted poultry: chickens, ducks, geese, the whole sheboodle! In part this is probably because both my grandparents and my best friends when I was six years old had chickens, and I must associate the fun times with them. The other, far more concrete reason was eggs! I imagined my hens happily laying enough for breakfast every morning, me skipping out merrily to collect them, then returning carefully without skipping to avoid the risk of breaking them!

As is my wont, the first step to buying chickens was to buy some books, three I think, which in all honesty told me very little. There was information about housing, but mostly aimed at those who were going to build their own chicken houses, whereas we had already decided to buy an ark. There was a small section about diseases and pests, a lot about chicken biology, and then some recipes for eggs (more than just scrambled, poached or fried). Finally there were many pages on the different breeds, some with full colour pictures.

For those who might be considering keeping chickens, I'll try and save you time and effort by summarizing the things I actually learnt from the books (unless of course you like pictures of pretty birds). Firstly, hens will lay eggs even if there is no cockerel (or rooster) present. This makes sense, really, if one considers a mammal's menstrual cycle, the only difference being that hens are on a solar (daily) cycle, and not a lunar one. (I shall resist some cheap gags about this meaning they're likely always to be grumpy in the afternoon...). An individual hen is unlikely to lay an egg every day, though this varies by breed, with some breeds getting close to that perfection, which is probably fairly exhausting for them! In addition the daylight cycles affect their laying, so they are likely to slow down or even stop over the winter months. This is why commercial egg farms prefer to keep their chickens in barns so they can use lights to keep the 'daylight' for longer during the winter; however, this wasn't something we were planning on worrying about, at least initially!

On food or drink the books were as useful as those on goats. Evidently chickens are thirsty birds and need a source of fresh and clean water at all times (though advice about any farm animal is unlikely to say they only need a trickle of dirty water ... but you never know). As regards food, amounts seemed to be about using rough handfuls, and as much as they'll eat in ten minutes but no more. So were we supposed to time them and then shoo them off after ten minutes? If they'd eaten some grass as well beforehand, did we need to take that off the time? And if they ate for eleven minutes would the sky fall in? As with all of these, the subtext was that we'd work it out for ourselves as we got used to them.

The final area of advice concerned cockerels. This was our first experience of a deviation to the usual rule about males in farming, in that cockerels are considered to have almost no use at all! There was much about 'sexing' chickens and eggs, and how the best thing to do is to get rid of the young cockerels early. The problem is that if one is keeping chickens for eggs, then cockerels present no actual benefit – in fact one of the leading authorities goes so far as to say that there is no place for them in chicken keeping at all unless you are intending to breed chickens. Others are more male friendly and say that a cockerel will tend to keep the flock happier, but that there should be no more than one cockerel for every three hens (clearly written by a cockerel!). Our own general inclination was to have a cockerel, if there was a chance he would keep the hens happy, and also the sound of a cock crowing in the morning is one of the quintessential signs of rural living!

Acquiring Our Chickens

Armed with our new-found knowledge we started on our quest to obtain some chickens. Our first step was to prepare their new home, so we ordered an ark from the only people who seem to sell such things, and although it was rather expensive, we told ourselves it was only right that we get the best possible housing for them. It was a triangular prism shape, with the house forming the top third and a run underneath. The website claimed it was big enough for twelve bantams or eight normal-sized chickens, and we were quite content with that as we were planning on four or five hens and a cockerel.

The ark was duly delivered and we were quite impressed with it: it had easy-carry handles and seemed quite robust. We placed it in with the goats on the slightly optimistic assumption that the combination of our capricious pets and the electric fence would discourage foxes, the bane of all chicken keepers.

Now we had to get the chickens themselves, which was actually more complicated than one might have imagined. For a start there is the major split between normal chickens and bantams, the latter being around half the size of a 'normal' chicken. Then within each of these groupings there are dozens of breeds, from the Light Sussex, to the Rhode Island Red and the Silkie, to the Wyandotte and Pekin. However, our initial plan ignored all these options entirely, as we had decided we were going to be knights on white horses (or maybe two knights on one particularly strong horse) and save some battery hens! We'd read about a lady who worked with battery farmers to rehouse their hens; she seemed to charge only 99p a hen as long as they went to a good home. The stories sounded terrible, though in the end heart-warming; evidently for the first few weeks we could expect the hen only to move around in a space roughly equivalent to the size of a sheet of A4 until it got used to having more space and relative freedom, though after a couple of months it would be clucking and scrabbling around like any normal hen.

We were sold, and Alex duly contacted the lady, and a couple of other organizations, and we were put down on a waiting list. But nothing happened. After a few weeks we became a bit restless, and then we read another article which said that the lady had been flooded with helpers and couldn't supply them all. Alas it seemed we would not be able to be knights, and I sadly hung up my spurs – though as far as I'm aware we're still on the waiting list…

After this initial setback Alex searched the internet and found a lady who was selling bantams not far from Portsmouth, so off we trekked one Saturday morning with hope in our hearts, and cash in our hands. Incidentally, Portsmouth isn't actually that close to us, being an hour or so away, even though it's in the same county. Some might wonder if we should stop using the internet, with its agnostic approach to distances, and look at local farm shops or ads, or even check the internet adverts for location – which would be a good point and well made.

The lady in question had quite an operation going in a field near her house, with several sets of chicken houses, a pond and ducks, and some geese. It all looked very impressive until we spotted one of the geese lying dead next to the fence – a fox had got to it overnight, through the fence, thereby instantly dispelling one of the myths I'd heard about geese being able to see off a fox. We selected half a dozen chickens, and the lady also tried to persuade us to take some ducks: they were Call ducks, a small but relatively pretty breed – her son had wanted them but they had bred like crazy and she really wanted to get rid of some of them!

We decided that ducks were a step too far at this stage, but asked her a bit about the geese, more from general interest than any particular desire to get any at that point. She informed us they were 'a bit of nightmare', that they churned up the ground 'something rotten', and was really rather negative about them. This put us off geese for quite a while.

The lady had told us that chickens are fairly easy to transport, we just needed some cardboard boxes, with perhaps some newspaper in the bottom, which we dutifully took with us. Then we'd catch the chickens and place them into the box, ensuring we closed it securely; they might challenge the roof a couple of times, but they would settle down and would be very little bother until we got them home. All very straightforward, or so we thought. She showed us how to catch them, by carefully covering their wings from above and behind so that they don't flap – though actually catching them was a little more complicated than this explanation would lead you to believe as they really aren't that excited about anyone getting too close to them. She also showed us how to hold them, by, in effect, getting them to sit on your palm with one leg between your forefinger and index finger, and the next leg between your ring finger and little finger, while also covering their back with the other hand to stop them flapping. I've also found it best to hold them close to my chest to give them some comfort and to prevent them from struggling too much.

With this knowledge, and some traditional wisdom from the lady on how much to feed them – a handful or two, not too much or they'll ignore it and it'll go off or attract vermin – we were completely ready to look after our new flock. This advice on food is typical of all animal keepers, and concurs with the information provided in books; nevertheless it was deeply frustrating as it really is no answer at all, so we planned to see how they responded at food times, and then adjust accordingly. Having chosen our new charges we watched as the lady placed them in the boxes, and then paid her a small fortune, before heading back to the barn, already anticipating some eggs!

We had selected six chickens, all bantams, and they were a fairly diverse bunch. Two were black frizzles, with crazy frizzly feathers that stick out all over the place and can either look like a stylish Royal Ascot hat or a complete mess! One of these was the cockerel, and we called the pair Sonny and Cher. Another was a lovely yellowy Pekin (or Cochin) bantam: these look like traditional chickens except are a bit more dumpy (apologies to them!) and have lovely feathered feet; we called ours Daisy. Then there were two others that were a sort of yellow and orange mix, still Pekins, which we called Luke and Beau (though they were hens ... here we go again...). The last of the group was a beautiful silver-laced Wyandotte – she had black feathers edged in white, and was simply stunning; we called her Bella Bella.

The books had told me that when newly arrived, chickens needed to be shut in their new home for a few nights to get used to it, which we duly did. The underside of their house had a chicken stairway which could be closed up to lock them in, and so we locked it for the first three days. We did open up the side of the ark in order to give them food and water.

I'd also decided that the run under the house was far too small, so I knocked up an extension run from some mesh and 2 × 4in wooden lengths. It wasn't the most elegant design, and it didn't actually quite fit, which necessitated some extra wooden bits, almost like bookends, to seal the join between the ark and the run, but it did more than double their space, and also allowed us to see them clearly. At this point we were very paranoid about foxes and weren't intending on letting them out of their run.

The Chickens Settle In

The chickens settled down and we soon established a daily pattern with them: we'd go out, feed the goats, then let the chickens out by dropping their stairway, and throw them some chicken feed. We'd check their water (which was in a chicken waterer, a well-named device), watch them for a bit, and then head in to get ready for work. In the evening we'd go out, feed the goats, and then put the chicken stairway up. After about a month of this we stopped moving the stairway on the basis that it was too much trouble, and they were sealed in anyway. In addition every few weeks we'd move the ark and run as the chickens would peck the grass and if we left it too long in one place the ground would be completely bare. It was during this process that we discovered the ark's weakness – the end doors. These were actually fairly flimsy, and sat on a flimsy bit of wood, which was soon broken. Eventually we resorted to putting a piece of chipboard against each end, held by bricks, to help keep the door shut.

One evening when we went out a magical offering was waiting for us: an egg! It was our first, and I can't really explain the sense of joy and pride I felt – it was almost as if I'd laid it myself! I immediately took it inside, put the frying pan on and cracked the egg open to cook it – and there had my second surprise. When I was a child I'd draw an egg with a really bright yellow yolk, one that was almost day-glow, but years of supermarket eggs had conditioned me to think that this was just youthful exuberance and that the natural colour of a yolk was a sort of pale and wan yellow. Well, the egg I cracked open had a sunshine-yellow yolk so bright it almost hurt my eyes! The yolk also seemed bigger than normal, and that, combined with the brightness of the yellow, made for a real 'Wow!' moment. And the flavour: it actually tasted of egg, the way eggs had tasted in my youth and which I'd also come to believe was an imaginary ideal of an egg!

From then on our lives have contained omelettes and scrambled eggs which are both properly yellow and have real flavour. I would recommend to everyone who has more than sixteen square feet of grass to get chickens – and if you have, and you haven't got any chickens yet, then go out right now and buy some! The reason the yolks are so yellow is apparently due to all the grass the chickens are eating (another gem garnered from the not altogether useless books). The grass is converted from green to yellow by the chickens, and clearly also adds a whole extra level of flavour. Soon enough we were getting three or four eggs a day, and even five a couple of times. I was doing my best to eat them all, much to Alex's chagrin: she said they contained a high level of cholesterol and were therefore bad for me. I tried to explain that it was a different type of cholesterol, but of course this was mere conjecture and therefore ignored by her.

Thus things were going very well for us with the chickens, with regular eggs, and no problems as long as we remembered to move their ark every few weeks to ensure they had fresh grass. We let them out occasionally so they could roam the field, though only if we were around, and they would range a little way from the ark, though rarely going more than a hundred feet or so. They were always very inquisitive, and if we had food would happily take it out of our hands. A number of articles I'd read had said that bantams tended to have a lot more character than normal-sized chickens, and ours certainly seemed full of personality. Cher was clearly top hen, strutting her stuff, though we were never sure if it was just because she was the same breed as Sonny or because she was the meanest pecker in town! Bella Bella, however, was bottom of the pecking order, and this was quite literally, in that all the other hens would peck her head if she got in their way, or pushed towards the food, or even got too close to them. This made her the most nervous of the bunch, and being so pretty she was wistfully reminiscent of Cinderella. . .

Bird Flu and its Implications

Then the bird flu scare struck. Suddenly all birds were a source of concern, the media went crazy, and smallholders like us were vilified as being a likely source of the bug (well, not actually us, but our equivalents in Asia who would, allegedly, have much poorer biosecurity), and if not a source, then a likely staging point as we weren't professionals. There was much discussion of biosecurity, and how free-range birds were bad because they'd happily mix with wild birds (I'm seeing a school dance here, with boys on one wall and girls on the other, but some of the free-range girls seem happier to move across the room. . .).

This particular part of the hysteria didn't last that long, but was irritating all the same, mostly due to the normal media desire to find someone to blame (or 'blamestorming', as it is often called). We kept our birds in their ark and run and didn't let them out again, and we also kept an eye on wild birds as much as we could to make sure they weren't having any illicit liaisons with our precious bantams. (Where exactly have you been young lady? You'd better not have been meeting Johnny or one of those other pigeons. . .)

Then about a month or two after the crisis blew up we came out one morning to find Sonny lying listless on the ground. He was still alive, but barely moving and not interested in food. Now Sonny was a lovely chicken; he had a strange little crow which never really got as far as a full 'cock-a-doodle-doo', but he would strut around his girls and often be first to peck the feed out of my hand; the Boney M song 'Sunny' would always be running through my head when I went out to see him. He was looking fairly bedraggled, but there were no marks on him, so we offered him some water, which he seemed to drink, and then some mashed food, which he pecked at a little but then ignored. We tried a few more things to get him eating and drinking, but he didn't seem at all interested. The next time we came out for another attempt (I think using a syringe, intending to spray water directly into his beak), he was dead. Suddenly we had a bit of a panic: here we were with a dead bird, and bird flu might already have got to our shores, so what if he'd died of bird flu?

Alex calmly decided that the sensible next step was to phone Defra and ask them what we should do. I thought they'd probably demand an autopsy or suchlike, which was a cost we didn't want to pay – but still, it was the 'Right Thing To Do'. But when she phoned she was met with cold indifference: they didn't care – didn't we know there was a bird flu crisis on at the moment which was keeping them awfully busy? – and when she asked what to do with poor Sonny, they suggested we bury him. Clearly we were no risk. . . or perhaps there was nothing in place to deal with small poultry keepers in such a circumstance, which should worry anyone thinking about the next possible outbreak. . . I'm not suggesting that they bring in a whole raft of regulations and movement forms, but at the very minimum some form of death checking service with autopsies where practical would seem a sensible precautionary measure?

We duly buried Sonny in the shade of a tree near to the ark, and gave him a short service. We also came to the conclusion that what had probably killed him was that he was completely exhausted! After all, he'd had five hens and not the recommended three, so he'd had a lot of extra work. . . When the flu panic finally subsided we started letting the hens out again, and we noticed that without Sonny there they never strayed more than a few yards away from the ark. They also seemed less happy. I realize I'm not supposed to anthropo-morphize them, but they were less clucky, less excited to see us, and some were so nervous they would now rarely take food from our hands. We thought perhaps it was the shock of him passing, but they were the same six weeks later – and so we resolved to get another cockerel.

A New Cockerel

In a little village near my in-laws there was a lady who also kept bantam chick-ens, and she had a spare cockerel that she was happy for us to have. When we went to visit she told us that she let the chickens 'get on with things', and sometimes one of them would turn up with a cluster of new chicks! She also didn't do the 'sensible' but obviously difficult job of killing off the excess males, so she had nearly as many cockerels as hens, and was more than happy for us to take one of them off her hands. She managed to catch one for us, out of a scrum of cockerels all apparently having a go at each other in a rather serious and potentially dangerous way. He was a (mostly) Light Sussex bantam, and a fine specimen thereof, with a white and grey body and black tail feathers.

We named him Lavender (we were both aware that it is rather harsh to name a boy Lavender, but that was the name which came to mind and that's what we stayed with – you'll have started to detect a rather wilful eccentricity in our naming by now...). He was full of life, and when we introduced him to the hens it was wonderful to watch. He'd come from a world where he was surrounded by other cockerels, mostly bigger, or tougher, or with more friends (allies?) than he had, and now, as he looked around, he couldn't see any other male. He seemed to check all the hens, and then look around, thinking to himself there was something wrong, and another cockerel might jump out at any time. When he finally realized it was just him and five girls, well, he was literally crowing with joy!

When chickens mate, the male jumps on to the back of the hen, grabbing with his claws, then bites the back of her head to gain purchase and does his stuff. If a cockerel likes one particular hen (or if you have a low ratio of hens to cockerels) then she tends to end up denuded of feathers on her back and head, as he takes one or two with him each time – probably not deliberately, but it could be a notch on the bedpost kind of thing. This is relevant because Lavender took quite a shine to Cher, and fairly soon she was looking rather bedraggled, with only a few scrappy feathers sticking out of her back. He seemed to like the other hens as well, and paid them all attention, but Cher was his favourite for quite a while.

So all was well once again with the chickens, and we settled back into a nice routine – until one morning we went to feed them and found the egg door open, and Lavender and Daisy missing. I maintain to this day that they had fallen in love and eloped to Ibiza and that we'd be getting a postcard any day soon, but another theory was that we'd been hit by a two-legged fox: a human thief. The main reason for this was the absence of a puff of feathers, and the fact that the other hens were alive and didn't seem particularly stressed (Cher perhaps even looked relieved...), and that the end of the ark with the nest box was open that morning (which could sometimes happen when we didn't close it securely and the goats knocked against it, but it was rare). Suffice it to say they were gone, and there wasn't much we could do about it. We thought about getting another cockerel but decided not to for a while as we had quite a lot going on with our other animals by this point, and the four hens seemed happy enough for the moment.

A Salutary Lesson

As a coda to the tale of our chickens, our neighbours Rob and Claire were at this point renovating the old farmhouse and had moved out to live in the village. However, they still had a group of sixteen hens at the farm which they would come over to feed every morning and evening, and to gather the eggs. I think they were Light Sussex; they were certainly white, and were 'proper' chickens, not bantams like ours.

Then one morning Rob came to feed them and couldn't find them in their shed – in fact all he found were fifteen puffs of feathers around the field, clearly where the fox had run them down and killed them. He'd lost all except one, which he found shivering in a hedge. It was a salutary lesson and one we took to heart, and it certainly prevented us from being complacent about shutting the chickens in when we'd let them out, and made us check round the run every day.

Fencing will Rule your Life!

Soon after we embarked on chicken keeping we discovered that chickens were no respecters of fencing unless it was mesh they couldn't get through or over; in this they were similar to the goats, albeit not really in the same league. We kept the electric fence going for the goats for some while, but felt that it was rather a small area for them, especially as it was in the middle of a much bigger field.

It was also clear to the goats that the grass was greener on the other side – and it actually really was, because, unlike the grass in their area, it wasn't being continually cropped low by them so grew up rather lushly. This seemed to taunt the goats considerably, and they would often contort themselves to reach just a little further under the fence, until in the end they would break through. We would then have to fix the fence again, and the routine would restart. I did try to give the fence a bit of relief by changing its shape, bringing some posts in so I could move others out, to reveal a few more feet of the lush grass.

There were three main reasons we hadn't immediately let the goats out into the large field completely: the first involved concerns about the footpath, the second was that we weren't confident of the fencing all the way around (with good reason!), and thirdly we were worried they'd gorge themselves on all the grass and end up suffering from bloat, or turn into giant goat balls that we'd have to roll around the field...

We started experimenting with mixing up the electric fence with the existing fixed fence to expand their area. I added some new permanent fencing where we planned to plant a hedge, and this formed one part of their barrier. We had an existing fence on to neighbouring woods, which formed another part of the enclosure, and the final part was electric fencing. The fence bordering the woods had three strands of barbed wire at about one foot, two feet and three feet off the ground, and had been put up by the farmer to keep in his cows; these were placid beasts and clearly unwilling to challenge a fence. I say this because it really wasn't that much of a fence: many of the posts weren't fixed in place, and the wire was pinned to several trees – though in fact these formed the most stable parts of the fence.

The goats, however, viewed it as no fence at all, and absolutely loved going into the woods as they were full of greenery, a positive Eden for them. They tended to squeeze under the lowest wire, but would also hop between the middle wires if the mood took them. The first few times it happened we managed to get them back in easily and tried to tighten up the existing fence, pinning it more, and trying to brace the posts. This didn't work.

Then they started going round to visit the builders next door, I think because they liked the company: they were always friendly, wanting to know what people were doing, and if it might involve food... At this point in time our neighbours' floor was being laid, which at that stage involving pouring the concrete; the main builder, a huge guy called Henry, would come round to report on the activities of the 'Escape Committee' as he called the goats on a couple of occasions. The most embarrassing time was when they had apparently wandered over and pooed into the just laid concrete... though I'm sure (fingers crossed) this won't affect the structural integrity of the building at all!

This predilection for escaping was starting to become a real issue, however, so Gordon decided to run an extra couple of lines of barbed wire along the fence using some spare wire he'd found in the main barn. He put one line about three inches off the floor and one at about one and a half feet – but to no avail: finally, said Bill and Ted, a challenge! And they immediately demonstrated that this was still no barrier at all, and we'd have people who were out walking their dogs coming in and asking us if the sheep in the woods were ours (very few people identified them as goats, as their shaggy coats really confused everyone).

I wasn't willing to put up with this, so added an extra line of barbed wire, basically along the line of the ground. There, thought I, they'll never get through that. However, if they'd been human Bill and Ted would have been limbo champions, because although it was clearly harder for them to get through, get through they did, and continued to eat as much of the woods as they could get down their gullets. Next door by this time had put up some temporary fencing which meant they weren't visited any more, but we were starting to worry that the people who looked after the woods might get upset and consider that we were actively allowing the goats to graze on their land. We were also worried that the goats would find something poisonous to eat in the forest and prematurely shuffle off this mortal coil, so we went for the next level of fencing – mesh.

We purchased some thick-wired mesh with holes of about three inches squared, which would be about four foot high when nailed up. We reasoned that the goats would not be able to get through such small holes, and if we nailed it close to the ground and bent it over a bit, they'd not be able to get under it, either. With these optimistic thoughts we set about putting it up.

This involved deploying one of the most useful devices I've ever purchased from ebay: a manual post rammer – a heavy, hollow metal cylinder covered over at one end and with a handle on each side, which could be brought down on a post to ram it into the ground. Ten to fifteen blows and the post would be firmly in, though it very often wouldn't be entirely vertical as it's actually rather tricky to bring down a heavy metal cylinder in a totally vertical manner. I have used it extensively over the years, and while I acknowledge that my fencing isn't the best, tending to look rather drunken on occasion, it does the job most of the time (depending on your preferred values for 'doing the job' and 'most of the time'). It is a little dangerous sometimes, and on one occasion Alex managed to ram her own head on the upstroke while trying to get a particularly recalcitrant post into the ground, resulting in a rather large bruise on her forehead and some mild taunting from me!

So using the post rammer, and treating it with respect, we rammed in some extra posts to replace those which were entirely missing or just broken, and used the trees where necessary, in effect creating a new fence over the old one. Sadly this didn't work either.

The main problem was that we didn't put the mesh under tension (as with the barbed wire) so it tended to sag in the middle (it was a few years before we learnt how to get the fence nice and taut). This meant that a committed goat could in effect climb over it, through the barbed wire, and be back rampaging through the woods in no time. We were at least partially successful as they didn't seem able to get under the fence, though this may have been because climbing over it was so easy. This led to some consternation on my part, until I finally (hopefully!) hit upon the answer and put another run of mesh above the first, tying it in (using the wire which had held the rolls of mesh together) as I went along, so the goats were greeted with a six to seven foot barrier.

Try as they might they couldn't get through this, where it was properly set up. Further along the fence there were places where trees jutted out, which made putting up the higher fence more tricky, but I thought the jutting branches, or heavy bramble in one case, would probably prevent further goat rampages. Mostly this was successful, but every month or two we'd either get a call from next door, or a walker would drop in to tell us our 'sheep' had gone walkabout. The goats were obviously tenaciously checking and testing the fence every day, and any fence would struggle to put up with that.

Each time we'd head into the woods with a saucepan of goat mix and a 'maaaaa' on our lips, catch them and drag them back. On occasion this included running madly around the woods, especially if Ted was feeling particularly frisky. We'd then walk along the fence and find the hole or sag and patch it up in some way or another. Fencing (and re-fencing) became a background theme to our lives from that time on...

The Magic Shop and Sourcing Supplies

A little later we discovered the 'Magic Shop', a farm supplies shop which was just down the road from us, actually a short walk across our neighbour's fields if we were so inclined. We called it the 'Magic Shop' because it seemed to have everything we needed: goat mix, chicken mix, sprays, signs, books and magazines (including the rather useful *Smallholder Monthly*), licks, posts, mesh, dog medicines, general vet medicines – in fact everything we'd ever wanted, and which we'd spent ages wandering from horse shop to garden centre trying to find, and all along it was two minutes down the road from us. We used to try and think of something it wouldn't have and then pitch up to find out if it did, and it always did.

After a while they began to recognize us, and we set up an account with them, and then the true magic was available – we could call them up and they'd deliver. Bales of hay – sure, will Tuesday do? Some more goat mix, not a problem. It was like rediscovering telephone pizza delivery (which unfortunately we didn't have, and even the Magic Shop did not provide), and rather sadly we were delighted by it.

I should explain a bit more. Buying supplies for a smallholding is like a couple choosing between a corner shop and a bulk-buy supermarket: at the corner shop the sizes are small and expensive; at the bulk-buy place everything comes in huge volumes, and is (for what you actually need) expensive – thirty litres of olive oil for £30 sounds good, but not if you're only going to use two litres of it. It's the same for smallholders: wormers always come in bulk, horse shops (which are often the only places with any real livestock supplies) only sell one or two bags of non-horse-related feed, licks come in blocks of two, which would last our goats about a century, and so on. Also the internet has not yet properly arrived for the farming and smallholding retail industries; it is possible to get information, forums, or details of the contents of the smallholder range of feeds, but it's very difficult to get anything which is going to be delivered, and the few places I've found have all been more expensive than our local horse place (which was already expensive, though I only realized just how much when we started paying more realistic prices at the Magic Shop).

So for several months we'd been stuck and starting to despair – and then as if by an act of sorcery, the Magic Shop appeared and took all our sourcing problems away!

EXPANDING THE GOAT HERD

In the summer of 2006, while we were having fun with the chickens and fencing, we were also worrying about the goats. Bill and Ted by this stage were about nine years old (though it wasn't entirely clear from the movement records), and ten years is a good age for a goat. We were worried that if one died the other might get lonely, and so came to the not unreasonable conclusion that we should find them some companions to ensure they didn't suffer from loneliness in their dotage. In retrospect I'm not sure that this was so reasonable, because in reality the goats were pets, they weren't providing us with food or anything else we could use or sell, and we should perhaps have considered that once they were gone it would be sad, but a chapter we could comfortably close. This rational approach never occurred to us at all, however, and so we rather blithely went about adding to our increasingly eccentric collection.

Unusually our first port of call was not a search on the web: instead Alex had excitedly noticed that there were quite a number of adverts for stock at the back of the smallholder magazine we subscribed to. One of my habits is to subscribe to any magazines covering whatever we're interested in, as I find they often have interesting information in them. Sadly most seem to repeat after about a year, so I then tend to cancel my subscription. There were quite a number of adverts, and they offered multifarious possibilities for increasing our stock; Alex often pointed to the adverts for piggies, of which there were usually many, but I always demurred (of which more later). Then the next issue came out, and one of the adverts was for some Angora goats on the Isle of Wight.

The lady was selling as she wanted to reduce her flock of goats which had been growing and was now too large for her. We wanted two female goats to join ours as we thought this would be the best balance. We'd also heard that billy goats were nasty, smelly and a real handful, and we felt we weren't ready for one yet (or perhaps ever!), or the related implications of kids. The lady confirmed that she did indeed have two females we could purchase – but she also offered to throw in a castrated male for free, if we were willing to take him, as they were a triplet and she didn't want to split them up. She assured us that he had a very sweet nature, and as he was castrated he was much easier to handle than any billy would be, or even some of the nannies. They were about two years old and had just been shorn, and she helpfully sent us a set of pictures.

They were a good-looking group of goats (in the sense of looking healthy and goat-like, and recently shorn, which always helps a goat telegenically), though it wasn't entirely clear which would be ours as there were nearly a dozen in the pictures. Still, it all looked good, so Alex agreed we would buy them, and arranged a delivery time.

The New Goats Arrive

In no time at all the new goats arrived, in a funky half-height trailer, just perfect for transporting a small number of goats or sheep, and we led them from the trailer to their new home. Unlike when Bill and Ted had arrived we were old hands now and knew exactly how to handle them, so grabbing their beards, which were not as long as Bill and Ted's, and offering them the saucepan of goat mix, we confidently led them into the field without any incident (Alex remembers this as a bit more of a struggle and that we dragged them by the horns – I prefer my version.) The lady was amazed that we were feeding ours goat mix: she considered it a real luxury, and asked us if they picked out the nice bits.

We'd never actually considered feeding them anything else: they were goats, and therefore we fed them goat mix. If they'd been elephants we'd have fed them elephant mix, and giraffes would have received giraffe mix (were such feeds available from our local feed shop, and ignoring all the other potential issues about keeping such creatures on pasture in Hampshire). This may seem naïve, but genuinely we assumed that goat mix would have all the right nutrients for them and we wouldn't have to worry about nutrition (which is a key consideration and usually takes up a lot of the how-to-keep-animals books).

The other part which amazed us was the idea that the goats would pick out the nice bits. Our goats seemed to be behave much like vacuums when it came to food, and with goat mix in particular, there was almost never anything left. It seemed odd to us that any goats would be picky, and it slightly dented my confidence that we understood how these creatures worked. I shouldn't have worried, however, as the new goats loved goat mix just as much as the old ones, and never left anything either. Looking back, I do wonder what that lady was feeding her goats. . .

So the new goats were happily in the pasture, and hadn't yet met Bill and Ted who were off in another part of the field engaging in a two-goat campaign to eat as many low-hanging leaves as possible. We sorted out all the paperwork, and exchanged some cash for a receipt, and the transaction was complete. Then in passing as she took her leave the lady mentioned that 'just that morning' she'd noticed some sort of light infection on the front of the male's legs, right above his hooves – but it was probably nothing and if we just sprayed it with antiseptic it would probably heal very quickly. She then set off home quite happily.

The goats were two years old and had just been shorn – and we found out later that this is the last valuable shearing from an Angora goat, because after that point the fleece becomes harder to work and is a lot less valuable – the sale price literally falls off a cliff. Secondly, there is no way that the 'infection' had only just happened: it was clearly something he'd had for a while – and he would have for a while yet. In summary, we were done like kippers. It taught us the value of asking more questions, and that if someone brings an animal with something suspect about it which they've 'only just noticed', the sensible thing is to just send it back with them! To be fair we probably wouldn't have sent the goats back, but we should have asked for a reduction. It reminded us always to remember: *caveat emptor*.

So three more goats were now with us; they already had names – the boy was called Bertie and it seemed to suit him, but we decided to rename the girls. One of them had a rip in her ear – clearly she'd pulled out a tag in a nasty way when she was younger – and we called her Belinda, not related to her ear but because she reminded us of the rather bossy chef who provided the (quite divine) food at our wedding. Really this was a compliment, even if the chef Belinda might not have seen it as such! The other girl we named Boris, I've no idea why, and I probably couldn't have explained at the time, but it was somehow right for her.

We introduced them to Bill and Ted expecting them to smell each other and perhaps nuzzle each other gently – but there was none of that. In fact it was immediately into pushing, shoving, head butting and all sorts of nasti-ness, much as had occurred between Bill and Ted when they were shorn, only now it was three on two (as both sides had already bonded); also the three new ones were younger and fitter, which made it harder on the original two. Even worse, we then sheared Bill and Ted the week after the arrival of the new three, which meant that not only did they have to contend with the three strangers we'd just introduced, but they also now had another partial 'stran-ger' to deal with.

We thought everything would settle down after a week or two, but it didn't. Even after three weeks we felt that the new goats were still 'picking' on our goats, which was silly but that's the way it seemed. Part of this I think was frustration with ourselves for being stitched up, which we unfairly projected on to the new goats.

After another week of this disruption I decided to split the goats back into two flocks, so I set up a separate electric fence about a hundred metres up the field. Following the logic that they were newer and therefore hadn't got used to the original area yet, we determined to move the three new goats to the new area. Evidently I had completely forgotten how hard it had been to move Bill and Ted from the old shed to the new, and was therefore strangely confident that I could split them up easily. Alex wasn't helping me as she was out, but I was sure it would be no trouble to move the goats on my own. Armed with a saucepan of goat mix I coaxed the three younger ones out of their existing pen, closing it behind them to stop Bill and Ted from following, and then walked slowly into the newly fenced space. It took me half a dozen tries to get them to follow me all the way, and eventually I had them in their new home, and closed the gate up.

Satisfied they were in, I started walking back to check on Bill and Ted. But suddenly there was a loud bleating, and next I was being mobbed by the three goats – then they were running past me and back to the original home part of the field, where they milled around a bit. They'd pushed straight through the electric fence! I checked it, and noted some areas where grass was touching, and decided that this was what was reducing its potency; so after dealing with this grass, and tightening up the wires, I tried to move them again...

I think I tried four times to move them, and it may well have been more, as the whole operation descended into farce. I quickly became exhausted with all the running back and forth, but stubbornly refused to stop. The goats seemed to love the whole activity, especially with the delights of goat mix. If I tried to drag them I'd manage to get one in, and then by the time I returned with a second, the first would be out. After several hours of this I gave up, changed tack, and decided to move Bill and Ted instead – surely this would be easier.

Sure enough they were definitely easier to move; the only problem was that once the other three saw I was moving them, they all galloped after me to join Bill and Ted in the new area. I was then in the challenging position of trying to move the three younger ones back to the original area, while keeping Bill and Ted in the new one, which they weren't exactly happy about. No matter which way I moved them, I realized that they'd formed a flock and I was trying to break it up, and it wasn't going to happen.

In some ways I think it was hard on Bill and Ted, and yet it did become easier to handle Ted after this – I never had to run around for fifteen minutes to catch her, she seemed much calmer and was always easy to grab, so perhaps we did do the right thing despite all the associated getting-to-know-each-other violence. The new goats took over the goat hut and would push the other two outside, even if it was cold and raining, so we set up a new shelter for the other two; but then Belinda would go and try and move them out of that, too. After a couple of weeks of this we settled with three covered areas, the shed and two lean-tos, and this seemed to provide enough that the shelter stress abated a little. But only a little...

A New Shed

As if it wasn't enough for the new goats to throw Bill and Ted out of their shed, they also then took to destroying it with wild abandon. Bertie in particular would spend hours battering it from the inside, and since it was only a cheap garden shed it could not stand such punishment for very long. One weekend we got up late and went out to feed the animals and were dumbstruck by the sight of the shed, which had almost completely collapsed – it looked as if a giant had pushed down on the top until the sides splayed out and it had slumped to the ground.

This was irritating, to say the least, and my view was that on no account were we going to buy the goats a new shed if all they would do was batter it to pieces. So we set about rebuilding their shelter using the bits of the old shed and some pallets that had been collecting in a pile with each delivery of building supplies for the barn (except every now and again all the pallets would disappear as one of the builders, or one of us – read Alex – would decide that a bonfire was in order). We also found some old tarpaulin to provide some form of weatherproofing. An hour's effort and their new home was ready: it had a central area with a floor plan the same size as the original shed, and two lean-tos big enough for two goats each, one on each side.

It was possibly the ugliest structure we'd ever been near; it was hard to clean, and it provided only just enough cover – but that was what the goats lived in for the next nine months or so. They actually seemed quite happy in their new home; I think it was cosier than the shed, and much more like the caves they might have occupied in the wild. There were a few more attempts to batter it, but it was fairly easy to fix each time. The strangest thing about this was that Bertie was never violent at any other time or in any other way: he just seemed to hate that shed. Belinda and Boris had given the original shed a little punishment but didn't seem to bother about the new structure.

My neighbours only mentioned the 'shed' in passing some months later to indicate that they thought it was rather ugly, and that it had attracted note in the village. Apparently people couldn't decide if we were being sensible by reusing the old shed pieces, or just cheap! Eventually the shelter began to look battered and ugly even to us, so we decided that perhaps it was time to do something about it and availed ourselves of another B&Q sale, where we managed to pick up another shed for a bargain.

We decided that Bertie would not be given the chance to destroy this shed. Our strategy was to build a strong internal frame of wood at goat battering height, and we also strengthened the floor by nailing some chipboard on it. We were very thorough and built something we thought the goats wouldn't be able to batter through no matter how hard they tried. Having completed our preparations we unlocked the door, strewed some straw on the floor, and stood back to watch.

The goats entered, peering around them, and then proceeding to eat the straw, provide some extra decoration (poo), before sitting down happily. Over the next few weeks it became clear that Belinda, Boris and Bertie considered this to be their new home. We spruced up a lean-to for Bill and Ted, and they seemed happy too. After all our efforts it would have been good to know that we could hold Bertie in, but he never once challenged the shed while we were around. It was almost as if the goats were telling us they didn't like the old shed, and once we'd bought them a new one there was no reason to destroy it as it now performed to their exacting standards!

The Flock Settles Down

Goats (according to some source or another) are always ruled by a 'flock queen'. With just the first two it had been Bill who was queen, and with the new three it was clearly Belinda: she was always the first to the food, and would push the others out of the way aggressively if they tried to interfere; she was also the most inquisitive, even more so than Bill. I think those first few weeks were toughest on Bill as she'd previously been queen and wasn't so happy about having to give it up; but eventually dominance was settled and the flock settled down.

Coping with Bertie's Infection

With the more relaxed atmosphere of the flock we had time to start worrying about other things, like the infection, or rash, or whatever it was on Bertie's legs. We tried spraying it regularly with antiseptic spray, but apart from giving him purple legs it didn't seem to be achieving anything at all; so, conscious that our 'free' goat was now going to cost us extra money, we sent for the vet, Emily. She arrived in due course and inspected Bertie's legs, and told me that it looked as if it was either an allergy or possibly mites. Her suggestion was that we try spraying it with antiseptic for a couple of weeks to see if that cleared it up. It didn't.

The next suggestion was that we use a special scrub to clean the area regularly (daily was suggested, but I think this was adapted to every few days). A small tub of pink hibiscus scrub was procured, and Bertie spent the next few weeks being regularly upended to have his lower legs washed vigorously; this left him slightly paranoid as he never knew when we next might pounce on him to aggressively clean his legs, and regularly looking rather natty with pink socks! This treatment had no effect on the rash either.

Plan C was to give him antibiotics for a week. These were to be injected into his rump (in one of the fleshy triangles between his hips, bottom and spine) every day, with the added hint that we might want to do this while he was eating. Strangely he seemed more bothered by us creeping up behind him than the actual injection. I'm still not happy doing these, as I'm concerned that I might not get it in the right place and cause complications, or that it will fail to work, so Alex was manoeuvred into doing most of them! A week of this, and then we waited a couple of weeks to see what impact it mght have had. Once again it had no effect at all.

It was clear that Emily was running out of ideas, and her final recommendation was that they take a biopsy, as that would tell us whether it was mites or an allergy, and then we should be able to specify a more targeted treatment. This sounded expensive and no fun, but after all this effort we were worried that it would get worse, and that it might affect the other goats, so we went for it.

Emily arrived with another more junior vet, with the quick explanation that as there weren't many goats around it would be good experience for her, and that an extra pair of hands would help. The 'experience' comment should have got me worried, but I was thinking more of how we were going to get the biopsy. This was in effect a cube of flesh from the affected area with the right size being about half a centimetre on a side, which is actually a fairly big chunk when you think about it. This was early in my days with the animals and I was still rather squeamish, which complicated matters somewhat as I really did not want to see the blood and gore.

I grabbed Bertie, which was always easy if there was some food offered, and flipped him on to his back, then tried to hold him as securely as possible, while also making sure I was not facing the point where the biopsy would be taken. Emily and her assistant went to work on poor Bertie, and after a longer period than I would have thought necessary they announced they had the first biopsy, and had sewn up the wound. However, they needed a second biopsy just to be sure, and that needed to be taken from a different leg, and unfortunately the best one was right in front of my face.

Bertie was fairly unhappy by this point, and I didn't want to let him go or move around too much for fear I might lose grip on him, so I was obliged to watch the cutting and sewing from fairly close up. Emily seemed to be taking such a large chunk of flesh that I feared he might have struggled to walk – but then it was all over and we could let him go. I gave him a handful of feed as a reward, and as he wandered off he seemed no worse for wear, and my squeamishness had been reduced a bit...

The biopsy was sent off to the lab, and we waited with bated breath for a result; after all, once we knew the answer we'd be able to get him sorted, and all our animals would be in perfect health. Emily soon called me with the result: after careful evaluation of the samples, the lab had concluded that the cause of the rash and scabbing was, wait for it … either mites, or an allergy! I expressed some surprise as to the lack of further detail, but was told that this was all there was.

It seems we don't know much about goats, but I felt that we didn't have to spend nearly two hundred pounds in order to further our ignorance. Behaving calmly, I asked about treatment, and was told that this would be much the same: we should clean the area and spray it regularly with antiseptic. My faith in the veterinary profession was somewhat shaken by this, particularly as we had already been doing this when we called them! After a few months the rash cleared up, whether due to our spraying it or for some other reason, we don't know – but it reinforced the message that we needed to be more careful about how we acquired our animals.

Boris in Trouble

Meanwhile life moved on apace, with summer moving into autumn, and we settled into a regular routine, feeding the animals each morning, and as the nights became longer, feeding them at night as well. This was to make up for the decreasing food value of the grass due to the lack of sunlight; we learned that once winter has set in properly there is almost nothing of value left in the grass and additional feeding is required. Occasionally we'd hear one of the goats bleating harshly as if it were in pain, and the first time it happened I rushed out, concerned that it was being attacked, or was caught in wire or something. Instead it just seemed to have lost the others, noticed their absence and was running towards them.

This happened several times with Boris, which led us to realize that she was in fact deaf. She might be grazing away at one end of the field, not facing the other goats, and if they all wandered off – bleating as they went – she'd not notice for quite a while. When eventually she did turn round they wouldn't be there, causing her quite a shock, and so she'd bleat madly, rushing towards them if she could see them, or back to the goat hut if she couldn't! However, she seemed absolutely fine otherwise, so we didn't worry too much about her – though we made sure she got food when we fed them, even if that meant going out to her in the field and tapping her on the shoulder!

Then one morning we were heading out to feed the animals and we could hear Boris bleating piteously from the field. This was fairly normal so we didn't pay a huge amount of attention to it, until we got out to the goats and four of them mobbed us as usual, but Boris seemed rooted to the spot about a hundred feet into the field. This was unusual in that the goats would always run to us when they heard us – or saw us, in Boris' case, but on this occasion although she could clearly see us, and was bleating away constantly, she wasn't moving. We quickly dumped all the food for the other four goats and ran to Boris.

At first we could see nothing wrong – initially I'd been worried that she was trapped in some electric wire or old fencing, but it wasn't that. She seemed to be favouring one of her hooves, so we set about turning her over to get a better look at it. Alex was nearest the offending hoof and peered at it, and realized that in between the two hard parts of the hoof was a small wound, out of which was wriggling a large white maggot. This really freaked us out, and we almost panicked – it was our first attack of fly strike. Fortunately Alex quickly identified the correct course of action, grabbing the maggot and pulling it out and then squishing it! I rushed into the barn to get some Crovect, which we applied liberally to Boris' foot, and another couple of maggots came out of the wound, but no more.

Once the horrid things were out, Boris seemed fine, and when we put her back on her feet she happily trotted off to get some food. As the other greedies had eaten it all we gave her some more, with the others cheerily joining in. We then showered and headed into work, in my case with a little blue dye on my hands.

Future Plans

The incident was an important one for us, and caused a robust debate about what we were actually doing, not just about the animals (though Alex pointed out that picking maggots out of a goat's hoof at six in the morning was not her idea of fun), but about the whole working in London and living in the country thing. As a result of this we became more committed to the animals, both in terms of wanting to do more to look after them (though how we could have protected Boris' hooves I don't know), and also in planning to get out of the Big Smoke and move to the country proper. However, there was one thing I was still sure of, and that was that I didn't want to be a farmer: it is hard work and wouldn't pay well enough to allow us to continue with the barn conversion, and I also think that when a hobby turns into a job it can destroy all the joy which has been found in the hobby. Nevertheless, we did know we wanted to live out in the country permanently and have land, and it gave us a context for our future plans, and also made us comfortable in adding more animals to our collection...

THE SMALLHOLDER'S COW

In the summer of our second year at the barn we heard a rather upsetting rumour – that the council was considering adding a number of houses to the village as a result of a government pledge to build millions more homes. Now, at risk of being considered insensitive, my contention was that the village was large enough, and we were very close to a largish town with plenty of housing – it was clearly one of those silly central targets. To be honest, however, those were minor dissentions, and the other side of the argument – that people need somewhere to live, and why shouldn't a few more houses be added to the village – also had some resonance; so we would have sighed and remained neutral, but for one thing: they had proposed one of our fields as a prime site. This was war! So Alex looked up the legislation to help us prepare our defence – and discovered that unfortunately there wasn't any, and that if the council really wanted to take the land they could force a compulsory purchase.

While not a legal consideration, one thing which would reduce the likelihood of this happening was if the land was being actively used. Unfortunately, at this point we were simply growing grass to sell to a friendly farmer, which isn't really an active use. We had plans to use it for cows, and since they are large creatures which clearly take up a lot of space we thought it might give the council pause in any attempts to appropriate our land. However, they really are rather big, and the thought of dealing with them made me nervous, which meant I kept putting off doing anything about it. To be honest we might never have got any cows, and the council might well have taken our land without us doing anything substantial about it, except that we had a major shock.

This was the death of one of my friends from college. He contracted a virulent form of cancer and died within a year of it being diagnosed, only a few weeks before his thirty-first birthday. Al was one of those people who would light up a room, always joking around, or telling amusing anecdotes or just generally being entertaining, yet underneath it all he was one of the brightest people we've ever met. His loss was a real shock to us all, it was so sudden: one year he was alive, bursting with vitality and bonhomie, and the next he was gone. Even years later I sometimes still can't believe it.

Choosing a Breed

Al's death made us realize that our time on this earth is precious, and that we shouldn't waste a second as it might be our last; we didn't want to look back and regret not making a decision just because we were being a bit wimpy (though as regards getting cows it was really only me being a wimp about it, Alex was rather sanguine). So we decided we really would get some cows, and I think Al would have appreciated the eccentric logic of the decision. We researched on the internet, and also, as usual, purchased a number of books: these, like those on the other animals, seemed to delight in contradicting each other, but they allowed us to deduce the following:

* While Highland cows are very cool, they are rather large (a visit to the local agricultural show confirmed this).
* Milking was not an option. This is important, in that dairy farmers do one of the least forgiving jobs there is: come rain or shine they have to milk their cows twice a day, every day. Cows don't take holidays and neither do dairy farmers, and that kind of regime really wasn't going to work for us. Also we didn't have any of the requisite machinery, which would mean that milking would be a manual and up-close process, which we really weren't up for!
* At least at first we wanted to start small, say only two to keep each other company.
* A bull was out of the question, for two reasons: first, we had a footpath running right across our two fields, and bulls aren't allowed in fields with footpaths (except at certain times); and second (and I feel I must add this out of honesty), they are scary, and at least to our minds, not for beginners.
* After looking through the various breeds we settled upon the Dexter, Britain's smallest native breed (about a quarter the size of a Highland cow), and according to several sources, an excellent smallholders' cow (I shall return to this point later).

With this comprehensive information (oh, how I wish we'd known!) we were ready to proceed to the next step, and find some cows. Alex duly scoured our off-net resources (*Smallholding Monthly*, I believe) and found what we needed: a nice lady called Philippa had two Dexters for sale, a cow and her heifer calf. She was selling them because her husband had recently died suddenly, and she didn't feel capable of looking after the two of them. In some ways the symmetry of this helped us make the decision, as we wanted to help someone else who'd recently lost someone. She invited us to visit and take a look at them, and this we did, though it was a trip of some 150 miles (once again, even with a non-internet search, we had failed miserably to take into account the whole distance thing).

We Meet the Dexter

Philippa lived in a cottage just outside a small village, surrounded by lovely Herefordshire countryside. The cows were being kept in an orchard field across the road from the cottage; the field was about two acres in size, with some stable-type farm buildings in one corner. We cautiously approached the cow, a black polled (which means lacking horns) Dexter of nearly seven years of age called Hazel, and she, not quite as cautiously, edged away. In fact we couldn't get closer than about ten feet before Hazel decided that that was close enough and moved away with more alacrity. Philippa explained that she didn't approach them too much, as she was a bit worried about their size, and so they weren't that accustomed to human contact. We didn't press the point, and surveyed the cows from a safe distance.

The mother was a fine-looking cow, though to be fair we knew very little about how cows should look. To us she seemed fit and healthy, with a shiny black coat and a firm gaze. We also managed to catch a glimpse of her calf, Catkin, who was only six weeks old and a lovely brown colour (though her sire was also black, so her grandsire's genes must have come out); Dexters can be brown or red, but black seems the most common colour. Catkin was also very cute. She was only about three feet tall, and stayed with her mum at all times.

Given that our plan was to release the cows into the field and let them be, with perhaps a vague plan to breed from them later on, we thought their nervousness should not be a problem. They were also lovely-looking animals, and small enough that it seemed likely we'd be able to handle them without a problem. So we said we'd take them, and agreed to call later to arrange a time to return and pick them up once we had organized a trailer, and to give her time to prepare the paperwork.

On the way home we determined that Hazel and Catkin did not seem appropriate names for our two imminent arrivals. In previous discussions we had alighted upon the idea of naming the cows after the seven deadly sins, under the dubious assumption that it would make eating them less of a cause of guilt. After some discussion we decided that Hazel would become Wrath, and Catkin would become Avarice. We didn't realize quite how prescient those names would be...

Preparing for the New Cows

Moving cows is a distinctly non-trivial process, and we couldn't just pick them up immediately; for a start we needed to resolve some paperwork with the authorities so we were registered to keep cattle. Also the cows needed to be tested for bovine TB before they could move, which for some reason, which wasn't explained at the time, seemed to take several weeks. We had no real clue as to what the tests were, so had no idea if this was a problem or not, though at this point we didn't worry about it too much, especially as this delay gave us time to prepare the new area for the cows.

I set up some electric fence to provide about a third of an acre of pasture around a corrugated iron shed in the centre of the field, which in days past had been shelter for some pigs. We spent a couple of hours opening up the entrance of the shed so it would be big enough to allow a cow in, and tidying up the loose bits of wood and metal which littered the area, until we were happy that it was a suitable home for our new additions. They would need access to more land in due course, but given the fencing which would be required we thought we'd start small, and then move on from there. At this point we also weren't confident in letting them share an area with the goats as we didn't know if they'd get on, or what the outcome might be if they didn't. A partially flattened goat would not have been a positive outcome.

As with all stock we needed to provide a fresh water supply for the cows. We wanted something temporary as we'd probably end up moving them, but we also had to get a long way out into the field and the thought of lugging buckets of water every day did not appeal, especially after our early experiences with the goats; so we bought some hosepipes and a water trough. We had to connect three long hosepipes to get out to the new cow area, and the hosepipe wouldn't stay in the trough easily, but at least we had something. We'd turn the tap on when we went out to feed them, and by the time we got to the trough the water would be flowing and we'd just have to stand and hold the hose while it filled up. It would be a little awkward if we forgot to turn the tap on, as we'd have to jog back and forth to turn it on, fill up the trough and then go back, but I figured we'd learn quickly!

Now everything was prepared and we were eagerly waiting for the green light to go and pick up our new stock.

Transportation Trauma

Some six weeks later we trekked off once again all the way to Herefordshire, though this time we had a horsebox and reinforcements in the shape of my in-laws, Gordon and Sue. On arrival we were met by Philippa and her neighbour, Peter, who was a dairy farmer. The two cows were safely shut up in their shed and had yet to be fed, which would hopefully make them easily biddable.

Under the farmer's direction Gordon carefully backed the trailer as close to the shed as possible, through the gates of a little fenced area within the larger field, and Peter set about building a barrier to guide the cows in, consisting of several sheets of wood, quite a few pallets and a lot of rope holding it all together. He spent considerable effort on it, and was muttering to himself while he was doing it. To be honest, in our naivety we couldn't understand why he was doing all of this – surely it was just a question of waving some food in front of the docile bovine and leading her in? Well, that's not entirely the way it worked out...

Peter finally declared himself ready, and standing in the corridor he had formed, he opened the door and stepped to one side. Wrath (fka Hazel) looked out and calmly stalked into the trailer, Peter following carefully behind. See, we all thought, that was easy! But alas, before Peter could attach the rear bar to hold her in, Wrath changed her mind and backed all the way out and back into her shed. Peter followed her and coaxed her out, and she emerged but clearly had no intention of going into the trailer again: after looking round wildly, she jumped over Gordon's shoulder.

It was so sudden it was almost unbelievable. It was also incredibly impressive, as not only did she clear over six foot five inches, she did so from an almost standing start. We were all in shock and just stared at her for a few seconds before flailing around trying to work out what to do. In the ensuing disarray the corridor opened up enough to allow Avarice (fka Catkin) to run out.

Fortunately the car and trailer blocked the way into the main field for Wrath, but in her panic to get away from us she tried to get through the hedge which formed one side of the fenced area, jumping and kicking as she did so. The hedge was behind a strand of barbed wire, and somehow she got over the wire and started to push at the hedge. At one point I thought she'd make it through it, but it managed to hold, and she swung away from it and back towards the field, catching herself in the barbed wire. She was straining at it and I was worried she'd cause herself a real injury. Not really knowing what to do I approached her, and discovered I could stop her pushing any further by the simple expedient of standing near her – despite (or perhaps because of) her previous leap, she clearly didn't want to come near me.

One side of the fenced area was a wooden crush, a sort of high-sided cage used to hold a cow still while she is examined by a vet or is treated for injury. These are often made of metal, but this one was home-made and consisted of stout beams on two sides, with a cross beam just below cow head height at one end and a wooden gate at the other. The farmer decided that the best thing to do was to shepherd Wrath into the crush and get a rope halter on her, so he opened up the back of the crush and stood on the open side to block her from moving out into the field, and I moved round and back so I was coming at her from behind. This seemed to do the trick, because she managed to get over the barbed wire to get away from me and miraculously went straight into the crush, where the farmer managed to close the back and hold her. In a further display of skill he deftly slipped the halter over her head and tied it to one of the posts, giving me the other end to hold.

Once the halter was on, and after a brief pause while she worked out what had happened, Wrath went crazy and started thrashing about, kicking so wildly that she destroyed the wooden cattle crush almost completely, knocking over the stout side beams and lifting the gate off its hinges. She didn't manage to loosen the post her harness was tied to, but instead somehow managed to get her head underneath the cross beam, and then wrapped back around the post at an awkward and clearly painful angle. This seemed to give her pause for thought, and she stopped kicking while she tried to work out what to do next.

Through all this I'd held on to the halter, more through shock than anything else, which is part of the reason she ended up in such an odd position – but at least she'd stopped kicking about for a bit. The farmer and I stood there looking at each other, both breathing heavily with all the effort.

This was all incredibly alarming and was causing some of our company to suggest discretion as the better part of valour. The others had all been carefully shepherding Avarice to ensure she didn't stray too far away, and Gordon apparently said to Alex that it might be sensible to give up and leave, but the message never got through to me. However, I'm not sure I'd have agreed to stop at this point anyway, as it was now a mission!

Gordon moved the trailer so that the back entrance was close to the post around which Wrath was wrapped, and the farmer attached some rope to the halter and also to a bar on the inside of the trailer. My memory of the next few minutes is fuzzy, but we effectively pulled Wrath into the trailer by taking up any slack she gave and slowly coaxing her up. In part I'm guessing because my memory is somewhat hazy, as at one point her hooves flew but inches from my face. It only really hit home how close it had been about two hours later when I thought more about it (and then stopped thinking about it!).

Despite all this we did manage to get her in and secured, but as we stepped back from the effort two things occurred to me: firstly, we had yet to get Avarice in, and secondly, if she carried on thrashing about like she was, she'd knock the whole trailer over. Alex said afterwards that the way it was rocking, it was more like the Incredible Hulk getting angry, and at one point she had thought that Wrath might just smash through the side!

To be fair, Alex, Sue and Philippa were not standing idly by while the men sorted out Wrath; they were covering the exits and had herded Avarice into one corner of the field. From there, getting Avarice in was much easier, with all of us slowly shepherding her back to the shed, where the farmer managed to get a halter on her, and he and I dragged her up into the trailer as well. For an animal barely three months old she was very strong and there were a couple of times I thought she'd get away, but we managed to hold on. Closing the trailer door was the seal on the victory, and if I hadn't been quite so exhausted by the whole run-around I might have danced in happiness.

A strong cup of tea helped to settle the nerves, and give us time for the exchange of cash and copious form filling. Some two hours after arriving, though it felt like several years, we were on our way back to Hampshire with all of us wondering precisely what we were to do with the wrathful creatures in the trailer... It was rather a long journey, and we'd set out early in the morning and missed lunch so we were all starving and decided to stop for a quick burger at a set of services. It occurred to us that we might be running a risk, with a potentially smouldering ball of rage in the back of the trailer, but we were very hungry!

We therefore went into the burger place and ordered our lunch, but every few minutes one of us would go outside to check the trailer was still there and in one piece. I think we had some half-formed idea that if the cows did get out we might be able to herd them back in, or something similar. Fortunately the trailer was solidly made, and Wrath seemed to have calmed down as there was no banging around (and indeed hadn't been when we pulled in). Still, it made lunch a somewhat syncopated and ultimately unsatisfying affair, and we were soon back on the road.

Wrath and Avarice Arrive

During that return journey we had plenty of time for the excitement of the cow loading to sink in. In fact it led to a long discussion, only interrupted by the talk of hunger which the burger had assuaged, on how exactly we were going to let the cows out, and who would open the trailer. Whenever we get new animals Alex and I tend to fight to be first to interact with them, and Sue is also always eager to get involved. For once, however, everyone was surprisingly reluctant to be the one who opened the door and persuaded the cows out. Through some age- and gender-related short straw it was decided that I was the one who would be sacrificed.

We had planned to back the trailer to the gateway in the electric fence in the naive assumption that they would daintily walk down into their new home, but instead Gordon decided that the best option was to drive the trailer right into the middle of their new area, and if necessary we could leave it there until they calmed down enough for us to drive out without problems.

Once we were in their part of the field I calmly lowered the back of the trailer, ready to jump aside at any moment if a wild-eyed bovine charged towards me. However nothing of the sort happened – in fact the two cows just stood at the back of the trailer looking at me, and even attempted to shuffle back a bit further when I bravely leant in closer. I suddenly realized that they, too, were quite shocked by the whole experience, and were also rather timid.

I decided to walk away from the trailer and out of sight so as not to discourage them from coming out, and we all moved a little further away from the trailer so our voices wouldn't bother them either. A few minutes later and they tentatively edged out of the trailer and into the field, and then they moved further into the field and away from the vehicle, keeping one eye on us at every step. Once they were a reasonable distance away from us we drove the trailer out of the electric area, and decided to leave them to explore their new home for the afternoon.

That evening we needed something a bit more fortifying than tea to wash away the stresses of the day. The farmer had said two things to us which stuck in my mind: one was that as a dairy farmer with a herd of over a hundred cows he would not have tolerated as feisty a cow as Wrath – she'd have been too much of a burden (meaning she'd have been for the chop – well, roast); and secondly, that when feeding the cows I should stroke them to let them know I was there and the one feeding them, and it would help them get used to me. Certainly having managed to get a handle on the goats (though I often wonder if it's not the other way around), I was not without some optimism that I could bring Wrath and Avarice round, and soon have them, quite literally, eating out of my hand.

The Perfect Smallholder's Cow

Now let us come to the matter of a Dexter being the perfect cow for the smallholder and the beginner. The main reason they are such a good smallholder's cow is that they are dual use, which means they provide both good beef and good milk. As regards the perfect beginner's cow, we were thinking small, docile and easily handled as being fairly important criteria. However, no sooner did we own two of them than we discovered that in fact Dexters are known as 'the small cow with the big character'.

Don't get me wrong, I like character, and indeed I have often myself been described as having plenty of it. However, I would not recommend myself as the ideal type of human, should someone want to start keeping humans, the main reason for this being 'character'. Thinking more on the topic I suspect that Dexters vary as much as anything else, and we were just blessed with two with extra character – and now that we have them, we wouldn't trade them in for anything.

Many people asked us what we were going to do with them, and whether we were going to eat them. My stock answer was that because Wrath was seven or eight years old we couldn't by law eat her (since the 'mad cow disease' outbreak in the early 1990s it was illegal to slaughter and eat cows over two years old, though I think that rule has now lapsed), and Avarice was far too 'cute' according to Alex. What we were going to do is breed them: if we had girl calves we would let them grow up and breed from them, and if we had boy calves we'd snip their bits off, fatten them up, and then eat them. However, at least at first we were just content to get them settled down.

The Cows Settle Down

As we'd done with the goats, we used to feed the cows in the morning, taking them a saucepan of goat mix (the saucepan was still our standard feed holder!). As ever the authorities were not entirely helpful on how much one should feed cows. The only practical guidance was that we should keep an eye on them, and if they were looking thin we should feed them more, and if they were looking fat we should feed them less ... which to be fair (and after some experience of our own) is not bad advice.

Wrath was very nervous with us and would approach only cautiously, and Avarice would not come near us at all. Still, we knew that if we carried on feeding them they would get used to us, and as the weeks went by Wrath certainly calmed down quite a bit. To an extent we also used this approach so that *we* could get happier with being close to the cows; they were, after all, still quite large beasts, despite being the smallest native cow. After the traumatic experience of picking them up it took a while for everyone to relax, but after a few weeks I did indeed have Wrath eating out of my hand. This can be quite painful as her tongue is very rough, and she was wont to snatch at my hand which can also hurt, but I learnt to pull my hand back at the last moment to avoid a set of crushed fingers. Nevertheless, progress was being made!

Avarice was still rather nervous, and being only three months old she was still surviving entirely upon her mother's milk (or at least so it seemed). When the two of them arrived we hadn't actually organized any specific food for them, so we fed them goat mix. After some initial scepticism she seemed to like it, and even came to love it, and so we didn't worry too much about getting some cow feed, figuring that with just two of them it wasn't worth the hassle of having multiple different types of feed.

After only about six weeks we decided that the area they were in wasn't big enough, partly because they seemed to have mown the grass down almost to the mud, and partly because we were sure they'd decided the shed was home, and that they would come to us if we called – and had the saucepan of course! So we removed the electric fencing and allowed them to have the whole field in which to roam, and they loved this and happily explored everywhere. Each morning I'd call them up (imagine hearing someone shouting 'Wrath! Avarice!' several times every morning at around 06:00 ... that was me!), and mostly they'd come to me, usually walking, if not placidly, at least without too much vigour – except on one memorable occasion.

There was nothing special about the day, it was just another work morning when I went out to feed the goats and the cows. As was often the case, the goats were right there and happy for food, and the cows were nowhere to be seen. I called for them and they didn't appear, so I started walking towards their hut – I was always punctilious about wanting to see them each morning to make sure they were still there and still well, even though some days I'd have to walk right to the bottom of the field.

On this particular occasion as I neared the hut Wrath appeared round the corner and raced towards me, 350kg of cow heading straight for me at very high speed. I knew that showing any fear to the cows would only make them more difficult to handle, and I think it was a combination of that, and shock, which kept me rooted to the spot as she sped towards me.

About six feet away she decided it was time to stop: she hit the brakes and came sliding towards me, literally stopping a few short inches away, with her head perfectly placed to get into the saucepan. My heart started beating again and I could breathe, and I stood there feeding her and stroking her head and feeling like I'd passed a big test. She never did that again, and I sometimes wonder what would have happened if I'd broken? Would she have run me down like the cowardly dog I was showing myself to be? Fortunately I'll never know.

The Next Project

Apart from these few teething troubles the addition of the cows to our little ensemble was fairly smooth, and in many ways easier than the animals that had arrived before them (though only if one ignores the trauma of picking them up)! We were definitely getting better at this animal lark, and with renewed vigour we looked forward to our next project: something we could actually eat! The friendly farmer renting our land had a number of definitely edible animals standing around in the fields, animals which seemed to need almost no care, and that helped us decide what we wanted next: sheep!

PART 2: GAINING CONFIDENCE

WE BECOME SHEEP FARMERS

Our first venture into sheep keeping seemed a good idea. The plan was to purchase seven just weaned lambs, fatten them up on our pasture and then have them converted into vacuum-packed chops and joints to last through to the end of the following spring. In order to avoid freezer overload (we estimated that each lamb should provide about twenty kilograms of meat) we were to share this bounty with my in-laws. Any costs were to be shared equally, including vets, food and slaughter, the key being that if one lamb was ill, for example, it would be treated communally with no attempt to pin the costs on any one party. This would also help in our depersonalization of the lambs!

We purchased the lambs from a friend who is a farmer, and he told us he'd bring our seven Texel-cross lambs as soon as they were weaned. We went for a commercial breed as it was our first experiment and we had no intention of breeding them. Also, to be honest, we had no real ideas of the differences between the breeds and he was the easiest person to get them from! I think he was happy to see us experimenting with farming – so many farmers are leaving the industry that I think they must sometimes feel like they're on a sinking ship – and as he's known Alex for many years I think he was hoping for some entertaining stories of how we were getting on!

In preparation we deployed a fence around half of our immature orchard as well as fencing the individual trees, and our plan was to release the lambs into this area. The intention was that they'd eat the grass around the trees (and hopefully the large clumps of nettles), and we could then move them to a different field at a later stage if they ran out of grass. The area was large enough to give them space to run around in, but small enough that we should be able to catch them fairly easily – and here I admit (once again) to some naive optimism. I'd connected up a water trough so they would have fresh water at all times, and with all the lovely grass we figured they had everything they needed. It was a key part of the plan to keep them separate from the goats and cows (who shared the main field) to avoid any fighting over food and other issues – particularly worms which we had been told can be easily passed on by sheep new to a holding.

Getting to Know the New Flock

When the sheep arrived they were a bit of a surprise. They were about three feet at the shoulder, probably already weighed twenty kilograms, and they certainly weren't the cute and easily handled creatures I'd had in mind! In fact they looked almost like fully grown sheep. I was assured this wasn't the case and they would grow quite a bit more over the next few months, though I'm not sure if this was to make me feel better that we'd get more meat off them, or silly for thinking they were full grown!

The very first evening Alex went out to give them some feed (well, goat mix, but we thought they would be happy with it), and say hello. As we had been looking after the goats for quite a time she thought she knew how they would react. Well, she wasn't in there for more than a few minutes, trying to coax them near and give them the food, when they all took fright and actually back-flipped over the fence into the goat and cow field. Now I think there are many people who would agree that Alex can be quite scary, but even at her fiercest I didn't think it would be enough to persuade them to Fosbury flop over a five-feet mesh fence!

The next day we resolved to get them back into the orchard. We don't have a sheepdog – though our two labrador pups would have loved to have given it a try – so it was just the two of us to try and carefully manoeuvre them back into the orchard. The goats would come to food, and that was our main method for managing them, but unfortunately these sheep had only been with us for a brief time and didn't know us as the source of lovely tasty food. In retrospect we probably should have at least tried with the food, but instead we went straight for the human sheepdog approach; this involved the pair of us spreading our arms out wide and gently encouraging them in the general direction of the orchard. We'd opened up a short section of the fence so they wouldn't need to demonstrate their athletic skills again, and the goats had rather helpfully decided to show the way by rushing into the orchard. Our first few herding attempts ended with the sheep slipping past us, and we decided we needed to be a little more aggressive in our arm fluttering, to try and get them to run into the orchard area. Anyone watching would have seen us running around our field like crazy people, waving our arms wildly and trying to encourage the sheep into the orchard, while also bickering as to the best way of controlling them!

After an hour or so of running round, slipping over and generally being humiliated by the lambs, we gave up. Lambs 1 – Humans 0. We left the orchard area open in the hope the sheep would go back in voluntarily (although I'm not sure what we thought would persuade them back, given their previous frightening experience!). This also meant the goats would be in there. We decided to try to stop the cows going in by putting a length of barbed wire, which generally I hate and try to avoid, across the gap about four foot from the ground, the theory being that it would prevent the cows but allow the sheep through. The main reason we didn't want Wrath and Avarice amongst the young trees was the risk they would eat all the leaves which could potentially kill the trees. It half worked, in that the cows didn't get into the orchard area, but unfortunately the goats did, and they were happily chomping away at the leaves, and even starting to eat the bark. We knew that if they ate a ring of bark all around the tree it would kill it, as the tree would no longer be able to get nutrients and sugars up and down its trunk.

Our solution was to get additional mesh and put it round each of the trees so that, in theory at least, the goats couldn't reach the trees from any angle. We were quick enough that none of the trees died, which was lucky, though several of them were much shorter after the experience! In the end we had to accept that the lambs would be out in the main field mixing with the goats and cows, though we still entertained some hope of getting them back into the orchard at some point.

Just a few days after we took delivery of the lambs the first outbreak of foot and mouth disease (FMD) was reported, and while we were outside the zones set up, we weren't that far away. The lambs had been brought to us from a farm which was further away from the outbreak than we were, but it was still a concern, and we watched them all even more carefully. This was relatively tricky at first, as they wouldn't come within twenty feet of us, especially after our abortive herding attempts.

Our plan was to slowly break down their resistance by offering them food every day. At first they wouldn't go near it, and so could not become addicted to its sweet pleasures, but they saw the goats swarm around the feed and eventually decided to investigate – and after their first few tentative tastings they became hooked! After a few weeks of supplying the good stuff to them they became much more used to us, and we could get within touching distance without them shying away. This meant that at the very least we could check that they looked healthy, even if we weren't quite at the control stage.

Initial Management Problems

They'd been with us for a month or so when we noticed that a couple of them seemed to have a slight limp, and resolved to keep an eye on them, and if possible grab them to investigate. It took another week before we had the opportunity to grab one of them, which involved me diving on to it while it was eating, which is rather mean, and holding on to it until Alex came and took proper hold of it, still struggling. Goats have an awful reputation for recalcitrance, but by comparison to sheep they are easy to handle – once you have hold of a goat they tend to be relatively calm, and flipping them over to do their hooves, while clearly not their favourite thing in the world, doesn't cause them to gyrate wildly.

Sheep are a different matter, however, and at this point we didn't know any way of handling this one other than to flip it on its back as we did with the goats – and it really did not like that. We eventually learned the proper way of handling sheep, where they are almost sitting up and leaning back into one's legs, which was altogether better as they are much calmer when in this position – though still not as biddable as a goat!

Once she was on her back we very quickly checked her hooves, and realized with some consternation that she had foot rot. We trimmed the offending hoof down as much as we could, but did not know what else we needed to do. Investigation on the web told us some very bad news, that foot rot is the result of bacteria which would now be living in our soil, and would probably transfer to our other animals as well. It was particularly likely to prosper in wet conditions, and we were heading into autumn...

The following week we also realized we would need to Crovect them again (they'd been treated before being brought to us) to maintain their protection from fly strike – our relatively recent experience with the maggots in Boris' foot made us extra sensitive to this risk. With goats this process is not entirely smooth, but made easier by the use of food, and the fact they have horns – Nature's way of providing useful control handles! With the lambs it was going to be a little more difficult, as they were a lot more excitable than the goats, and lacked these useful handles. This was of concern, as we had to get the Crovect spray all along their sides without getting any in their eyes or on their skin.

Of course in order to get to the point of spraying them I'd need to catch them again, and this time I wasn't going to be so easily defeated; so I built up a corral area with a wide mouth which we could shut, with the intention of encouraging the sheep into it. Alas, I again made a schoolboy error in that I failed to try persuading them in with food at the beginning. Alex was away that weekend, but two friends were staying and agreed to help... thereby leading to the spectacle of three people running crazily around the field, flailing their arms, with as little success as the previous attempt with just two of us! The sheep failed to go near the corral, and were then dismissive of any attempts to persuade them in with food. Lambs 2 – Humans 0. The only good side to this was all the extra exercise we were getting!

Fortunately our friendly farmer offered to help Crovect them (in part because he felt sorry for us, and in part because he felt guilty about giving us lambs with foot rot!), and brought along his dogs to round them up and some hurdles to contain them. Despite all this, it still took him four attempts, and he was close to giving up, which made us feel much better about our own efforts. We'll call that a draw: Lambs 3 – Humans 1. But at least they were now covered in Crovect and hopefully would have little risk of a fly-strike attack. He also trimmed their hooves, which helped to reduce the foot rot.

The weeks then passed somewhat easily, the main worry continuing to be the foot rot problem, which had also attacked our goats, causing them a lot of discomfort. Getting rid of foot rot is very difficult, and rather unpleasant; one of the best ways is to trim the rotten bit of the hoof until it bleeds and then spray it with an antibiotic, which clearly is no fun to do, and nor do the goats particularly enjoy it. However, by doing this every few weeks we eventually (after about four months) managed to clear it up. It does still come back occasionally, but fortunately is never as bad as that first time. It was also a lesson in bio-safety, and confirmed that keeping the lambs quarantined was a good idea, even if it didn't quite work out!

Lambs to Slaughter

The lambs seemed well enough, but constantly in the background was the worry about FMD. The implications of this disease become much more real when you have animals which may have to be slaughtered if it gets too close, or even worse, if they actually get it. By the end of September they were clearly no longer lambs and we were starting to wonder when they would be ready. We asked our farmer friend who had a look at them and said they probably needed a month or so more, and some additional feed. Up to that point they had been living on grass, with a small amount of goat mix in the morning and evening.

Feeding all the animals together was becoming fraught with hazard. With seven sheep, five goats and two cows all in the same area, there was considerable pushing and shoving – and not just from the animals: a couple of times we had to have words with the cow, as well as one of the goats, which kept gouging me with one of its horns. But the sheep were the worst – they'd mill around, preventing me from moving towards the feeding area, and on a couple of occasions one of them was so excited at seeing me coming with the food that she ran right between my legs, almost upending me. In addition, trying to give a few of them more feed than the others was virtually impossible, so they were all getting extra rations. Still, it's always nice to be greeted in the morning and evening by fourteen loving creatures. Okay, it was actually because they loved the food, but I could pretend they had some affection for me too!

When the shearer came to do our goats he said he thought the sheep looked ready, and so I decided to arrange the 'killing process', as my wife so delightfully calls it. We had agreed right at the beginning that she would have no part in this, so I had to organize it. Our original intention had been to have the lambs killed on site and then taken away, as this would vastly lessen the stress involved for them, but unfortunately we really don't have the facilities for this (though I intend to get them) so had to find an abattoir which would take them. We were advised to contact Laverstoke Park, a bio-organic farm with a state-of-the-art abattoir, as their practice was to accept other people's animals on one day each week, a Thursday. (They have since become extremely successful and you can buy Laverstoke products at Waitrose and Sainsbury's – I especially recommend the buffalo burgers!) I phoned them up and they were extremely friendly so I booked in the lambs for the following week.

In preparation for the big day I moved the lambs into the orchard by the simple expedient of feeding them in there. Well, it was simple for five of them, the other two I had to man-handle over the fence, but they didn't mind as they came straight down on top of the feed. Over the weeks they had become much more biddable, especially if goat mix was part of the enticement. I then closed off the orchard fence so they remained in the orchard area, and while they nosed at the fence a bit they didn't seem too upset, and they certainly didn't break out again. Around this time we also acquired two new Soay lambs, which we intended to keep as breeding ewes – but more about them a bit later.

Finally the fateful day arrived. My father-in-law had backed up our borrowed trailer to the entrance to the orchard, and we put up pallets to guide the sheep in. We had factored in thirty minutes to get them in, as I was optimistic that I had a handle on moving them. ... I therefore went into the orchard with our traditional food carrier – the saucepan of course – and walked carefully towards the sheep. In fact I needn't have bothered to be so careful because they ran up to mob me, so I turned around and headed into the trailer: the sheep happily followed me, even with a little pushing and shoving. I upended the saucepan into the back of the trailer and they dived on to the food, whereupon quickly and quietly I backed out, closed the trailer door – and we were done! It had taken all of two minutes to persuade them into the trailer and close it up: what a difference a few months of getting used to us had made!

We made our way to Laverstoke Park, and found the facilities. These were excellent, with a convenient bay to back the trailer up to, and a whole set of pens ready for the animals. There must have been at least fifty sheep in several different pens already there, but they all seemed quite calm and there was no plaintive bleating. Moving the sheep into their pen was relatively painless once they'd made the first step, and after filling out the paperwork and cleaning out the trailer with the broom and powerjet so thoughtfully provided, we were on our way home. It seemed almost too easy.

Laverstoke Park hung our meat for an extra week, and then cut the carcases into accepted retail cuts for us. At the time our only freezer was the small one in the fridge unit, and with at least four lambs' worth of meat due to come back to us – the other three were going to my in-laws – we realized that we needed to get a new freezer, and quickly! I did some careful calculations and found that we could afford to get one of the smaller chest freezers, and so we duly went out to purchase it. In retrospect I lacked some ambition here, as any success in the venture would encourage more sheep rearing and necessitate further freezing facilities, a situation which would be further exacerbated if the vegetables I was growing really got going – but it's all part of the learning experience.

When we picked up the meat it filled up the freezer completely! The meat itself was tasty, although one of the carcases seemed rather fatty – so we resolved that next time we would need to give the lambs a bit more exercise. Given all the running around *we'd* had to do, this was a little worrying, the risk being that we might exhaust ourselves before making any progress in sheep fitness training! However, I suspected the dogs would be more than happy to help us with this, though much more training would be required, and they might get a bit too excited.

All in all this had been a very successful endeavour, and we felt good about how we'd been able to handle the lambs, and that we'd been able to get some good meat out of the process. Apart from eggs, this was the first 'real' food we'd had from our animals, and the first bit of 'real' farming we'd done. Alex confessed that she thought I'd bottle the trip to the abattoir, but in fact it hadn't concerned me at all – I had always been clear in my mind what the lambs were for, and I never lost that confidence.

One partial mistake was that my sister-in-law had originally intended to take one of the lambs – before she moved to Bangkok – and had decided to name it Clarissa. She didn't select any particular lamb, just decided that hers would have that name. For some time after, whenever we served lamb we'd say that it was probably Clarissa... I think some people might have been rather put off by that, but fortunately the meat tasted so good that no one ever actually complained!

Two Soay Ewe Lambs

As mentioned before, just before we took the seven Texel crosses 'on holiday' (they love it so much they never come back...) we added two more lambs to the holding: Soay ewe lambs, which we named Lafite and Mouton (after the Rothschild wines: I was just in one of those moods). We were given them by the parents of one of my college friends: they have a small flock – a ram and three ewes normally – and each year they try to find homes for the ewe lambs, and the boy lambs go into the freezer. This year they'd had ewe lamb twins, and we headed off to Somerset to pick them up.

Before our trip we'd gathered some information on the Soay from the web and our books, although there is a lot less information on them than the main breeds. We were told that Soays are beautiful sheep that the uninitiated sometimes mistake for goats. They are generally a dark brown with lighter markings around the rump, and they usually have quite long horns. They are a much smaller sheep than the normal large woolly commercial breeds, generally half to two-thirds their size.

There is much to recommend the Soay: they don't really need shearing as their fleeces moult off naturally, they are less likely to be hit by fly strike, possibly because of their fine summer coat, and their meat (we were told) tastes better than normal lamb – a touch gamey perhaps, but certainly more flavoursome. On the down side they have quite a bit of character (oh dear), they obviously yield less meat, and they tend to be a lot less biddable than 'normal' sheep, having a weak flocking instinct and a distinct sense of independence!

Picking up the ewes was an interesting experience, as it was our first chance to see these animals at close quarters, and to watch them in action. Roger, the owner, had corralled his flock in a small area of their usual field around their shelter; the plan was to give them some food, and to grab the two ewe lambs while they were eating. We were instructed to stay behind until he'd caught them, and then to run in and take the lambs, and move them into the trailer. What I'd like to say is that it all went completely smoothly and we loaded the ewes into the trailer easily – and perhaps that's the way these things work for other people, but for us it never seems to be the case, and so it was with the Soays.

Roger approached carefully, with some food held out, but the canny creatures must have seen us in the distance (despite us hiding behind some trees) and were a little nervous. When he tried to reach for the first ewe lamb they all took flight, half of them pushing through the gate he'd come through into another small area of the field. So he then had to chase after them – but one rugby-like tackle later he had a writhing ball of brown fury in his hands, which he passed over to Alex, who had run up as soon as she saw that he'd caught one. Alex cautiously held the animal close to her, whispering soothingly, and it seemed to calm down a little.

Off went Roger to get the second one: this lamb didn't need quite the dive to the ground of the first one, but it wasn't far off, and another furious little brown creature was duly handed over – to me! Following Alex's example I very calmly held her close to my chest and told her that all would be well, and she was going to a new place where there were large fields for her to roam over. This seemed to work – but before we could put them in the trailer we needed to tag them.

I'd never seen this done before, as all our other animals had arrived already tagged. In essence a tag is a plastic loop with a sharp staple at one end which goes through the ear and fits into a hole in the other end, on the other side of the ear. A special hand-held device, like a sort of fancy pliers, is used to drive the plastic staple through the ear and into the hole at the other end of the tag. This procedure is probably as painful as having an ear pierced, though having never experianced this I really don't know what that level of pain might be.

However, it was clear from the way the ewes reacted that they didn't appreciate it at all! But interestingly they quickly subsided, and only seemed to shake their heads a bit to test the tags for a little while before accepting them, somewhat fatalistically to my mind. With that done, the little brown monsters were finally loaded in the back of the trailer, and having completed the usual forms, we were off!

When we first let them out the two ewes were very flighty, but judicious use of goat mix eventually had Mouton eating out of my hand every day. Lafite would sometimes eat out of my hand, but she was far more suspicious and was always quick to move away from me. They also seemed to calm down quite a bit when the other lambs went on holiday – I suspect there had been the usual fight for dominance and they were having a tougher time of it, given the size difference.

I'd like to claim that getting the two ewes was part of a grand plan, but to be honest it was more about the opportunity at the time, so we needed to consider what to do next. In the end I decided they were too young to breed from – as one farmer said to us, would you want a thirteen-year-old girl to get pregnant? These young ewes were the same, in that it was technically possible, but not really desirable. We therefore decided to get another Soay ewe in the spring, and perhaps a ram, or maybe we would use AI (artificial insemination). Clearly my confidence was up and I was starting to think of making sheep a permanent and sustainable part of our endeavour.

The Soays are Lost

Then one day we went out to feed them and they weren't there – they had completely disappeared. We searched all round the fields and couldn't find them, though we did find some weak spots in the fencing which we resolved to fix once we found the sheep again. We then searched the next-door woods and fields, and found nothing. We were getting a little panicky at this point, not just because they might be hurt or injured, but (again…) what if they got out on to the road and caused an accident?

Alex phoned the police to report them missing, and the policewoman duly took down her details. I thought this was unlikely to yield much, but at least we'd told someone in authority – and then she had a call from the police to say that someone had spotted some loose Soays, and they gave her a contact number. Blessed relief, we thought, as Alex phoned … but it turned out that it was some thirty miles down the road, and there were half a dozen of them which had managed to get into a farmer's field – but they definitely weren't his, or ours! He offered that if we caught them we could keep them … however, we weren't sure as to the legality of such action, or even whether we could catch them at all! I imagined us running round and round the field for some time before giving up and leaving with our tails between our legs!

After that there were no more leads, and we fell into a melancholy routine. Each morning we'd go out to feed the animals, hoping to see the Soays back, and then if we had time we'd trawl the nearby fields. After about three weeks we gave up, assuming they'd either died or somehow found a new home – but the thought of having to explain this to my friend and his parents was rather forbidding, and we kept putting it off.

Then one morning our neighbour Claire came back from a ride out on her horse to tell us that she'd seen our Soays mixed in with another large flock in one of their outer fields. We were overjoyed, and immediately set out to rescue (recapture?) our lost Soays. We took some goat mix and some rope to act as halters, and intrepidly set off across the fields. About two hours later we dejectedly trudged home again, covered in mud and poo and bruises, and having failed to get even the slightest touch on either of the lambs. We'd run around, disturbing the whole flock of about a hundred sheep, until we could run no more.

By some stroke of luck we then found out that these sheep were part of a flock that was due to graze on our land as part of a deal with our friendly farmer. The other part was that his farmworkers would put up some new fencing for us – this should have been the previous year but it had not yet been done… Anyway, the shepherdess assured us that when she took the flock off our land she'd leave our two Soays – and she duly did! While this had been going on we'd reinforced our fencing, fixing the various holes, and also tightening up the gates. It probably wasn't totally sheep-proof, but we had hopes! It also helped that all the neighbouring flocks had moved on, and weren't therefore sending attractive bleating noises wafting over our land.

After their adventure Mouton and Lafite seemed to get used to us again quite quickly, though for a while I checked on them regularly, both to make sure they were still there, and to see if they were pregnant. There had apparently been rams in with the flock at some point, and I was assured by the shepherdess that if our two had been in with the flock then, they'd have definitely been covered. I still don't know if they were or not, but it is quite common for ewe lambs not to conceive in their first year. The usual rule of thumb is around a 50 per cent lambing rate for hoggs (year-old lambs, and usually singles), but fortunately neither of ours was pregnant, as it might have led to complications (in terms of actually giving birth to lambs, due to their small size – as opposed to visiting rights for the father and the like).

We felt that the underlying reason they'd decided to go off on a frolic was that two sheep do not a flock make – so we determined that we would need to increase their number. Surprisingly the thought of getting rid of our two never crossed our minds, which shows how attached we'd become to them, even if it wasn't yet mutual!

Another Loss: Bill and Ted

Not long after this incident we went away on holiday to visit some friends in Hong Kong; while we were away we'd arranged for several different people to look after the animals, with our neighbours acting as glue between them as the dates didn't quite match up. We were having a great time when we had a call from our friend Molly who was looking after our animals, to say that sadly Bill had died. She was distraught as she felt she'd let us down, but there was nothing anyone could have done: it seemed it was just her time.

We were both really upset about the loss of Bill: not only was she the first of our big animals, but she was the first to have a real personality and teach us that goats could be fun. We'd become really attached to her, and it seemed unfair that we hadn't been there for her in her final days. Of course we knew that she was old, having reached around ten years old by this point, and a goat's normal life expectancy is between eight and ten years, but it was still upsetting. Phrases such as 'Well, she had a good innings' and suchlike rolled off the tongue, but it didn't make it any better; it was as if a favoured maiden aunt had gone up to the great tea room in the sky, there to knit, or in Bill's case, chomp happily on an abundance of ash leaves!

We stayed in Bangkok for the remaining few days of our trip as there didn't seem any point in rushing back, but Bill's demise took a lot of the shine off the holiday. When we flew back to London my in-laws collected us at the airport, only to tell us that Ted, too, had died. Well, losing Bill was bad enough, but Ted as well was almost a blow too far, and we were very subdued as we returned to the barn. We felt guilty for taking our holiday and leaving the animals behind – though clearly there was nothing that anyone could have done, it was simply because of their age.

Later we learned that it is very common amongst goats that have lived together all their lives that if one dies, the other will also die within a week or so. So I guess Ted just didn't want to go on without Bill. While we were unhappy, after a while we both realized that it was part of the cycle of life, and that if we were going to have animals, then we were going to have to accept it, and at least they'd had a good last few years as opposed to ending up as curry!

This loss, and the (fortunately) temporary loss of the Soays, drained us of much of our enthusiasm (shown, for example, by the fact that I stopped writing this book for quite a long period), and it took a few months of looking after the sheep, goats, cows and chickens before our confidence returned and we could take unalloyed joy in them all.

Eventually, however, we undertook another of our periodic animal expansions, this time in two parts: one was to find some companions for Lafite and Mouton, and the other was an entirely new animal for us – piggies!

BIRTHDAY PIGGIES!

Alex's birthday is in early June and I have a poor record of buying her presents. For some reason my mental faculties shut down and I either fail to buy her anything at all (once) or something that she really doesn't like (at least twice that she's admitted), so in our fourth year at the barn I resolved to try harder and get her something she really wanted: pigs. To tell the story of the pigs we need to step back in time a bit...

Early the previous year I'd become excited about the possibility of pigs. In the words of Homer Simpson (of the renowned television programme), a pig is 'a wonderful, magical animal' – or as I myself might say, 'I love bacon, I love pork, and that's all that needs to be discussed'. I'd duly done my 'usual' and searched the internet to discover what was required, as well as buying some books. These all clearly set out the requirement for shelter, amongst other things, so again using the internet, I found and purchased a pig ark, very much as a kind of statement of intent. This rather robust contraption looks like a small low hut with a peaked roof, and this particular type weighed an absolute ton, which made manoeuvring it rather challenging.

Around this time a friend gave me the book *Any Fool can be a Pig Farmer*, which I recommend as a great read, and very informative. However, it told me that pigs could be a real handful, and it did rather put me off. I particularly didn't like the story of the sow eating her piglets as they were born, or the description of sows killing their progeny by rolling over on to them. Pig lice also didn't sound like a lot of fun. So here we were with a pig ark and a suddenly less enthusiastic potential pig farmer...

Alex in the meantime had found someone who sold Oxford Sandy and Black pigs (Alex liked the name), and organized to pick up two little pigs. Unfortunately the weekend we were planning to pick up the pigs coincided with me being called away on a short-notice business trip, so we were unable to go. Alex phoned the woman to explain, and to try and rearrange, and the woman became really angry and shouted at her and told her it was an outrage and asked how she could treat people so appallingly.

The result of the conversation was that we definitely were not going to get our pigs from that woman. It also knocked some of our enthusiasm and we became distracted with other things, and the pig project went on hold. This was probably a good idea as we really didn't know enough about pigs to understand what we were getting into.

Over a year later I was starting to feel more comfortable again with the idea of animals in general, and pigs in particular. Pig farmers can be grouped into three main categories, no matter what breed of pig they farm (or so it seems to me): breeders, finishers and 'everythingers'. Breeders have sows and a boar or two, and concentrate on producing large litters of piglets which they then look after until they wean them, when they sell them on to finishers. Finishers buy in 'weaners', which are usually around eight weeks old or older, occasionally younger, and feed them up to the desired weight, whereupon they take them off to a nice man who converts them into meat, sausages, and perhaps bacon (if he has the facilities to cure, that is). 'Everythingers' do both!

There are arguments for all three regimes, and I'm sure there are many happy people in each camp, but the easiest to do is to finish. It used to be very popular for a family to get a couple of pigs to fatten up in the run-up to winter, usually on slops from the kitchen, to be slaughtered around Christmas and cured so they'd last through to the spring. Feeding them slops is unfortunately now illegal as a result of the furore around feed during the Mad Cow Disease crisis, and has in part contributed to the fading of this tradition. The reason it would normally be a couple of pigs is that it's never good to get a singleton as pigs are very social animals and can sicken if lonely, and even die in some cases. So the plan was to try out finishing a couple of pigs, and decide if we wanted to do more.

The breed of pig was an important consideration. We were more interested in one of the rare breeds, rather than a commercial variety, on the basis that these were reputed to provide the best meat. In our earlier phase of pig euphoria we'd discussed several breeds and Alex had made it clear that her preference was for the Oxford Sandy and Blacks, hence our early attempt to obtain some. These pigs are a sort of pink sandy colour with black patches on them, and according to at least one review I read of them they would be easy to keep, and an excellent beginners' pig. As I was buying them as a present I couldn't consult her on the breed further, but decided that she was unlikely to have changed her mind, so I set about finding some OSBs (as they are known to aficionados), preferably weaners.

The internet was once again my ally in this search, and in only a short time I had found two potential vendors – unfortunately the nearest one was in Kent, not that far from Dover, but a fair trek for us, and the next nearest was in Gloucestershire. Clearly the OSB really *was* a rare breed, and I realized that if I was going to get them in time for Alex's birthday I needed to get moving. I duly emailed the farm in Kent, but after a few hours still hadn't received a reply.

I must explain that at work I am on email constantly, and any delay in a reply is simply not acceptable, even if the reply is to say that someone is finding more information... I started thinking maybe they were away, on holiday, the email address was wrong, maybe their connection was down... I had to force myself to consider that in reality it was more likely that the farmer was not sitting near their email all day, but perhaps wading through pig poo, or planting feed or some such other activity. Anyway, about two days (or two eons by the standards of the wired generation) later I had a response saying that they did indeed have weaners, and how many did I want, and when would I be able to pick them up.

How many was a tricky question: clearly it had to be more than one, but did I want to risk more? I definitely felt two would provide the right amount of pork for us, so I decided to go for two, until Gordon said that if we were willing to look after a pig then he'd have one too, and pay a part of all the bills and so on. This seemed reasonable enough to me, so I asked for three (it also helped that three is Alex's favourite number) and arranged to pick them up the following week. The nice lady wanted me to pick them up as soon as possible as they were already weaner size, so I would in fact be picking them up earlier than planned, but in plenty of time for Alex's birthday!

The Weaners Settle In

Gordon drove me in his venerable old BMW estate to the wilds of Kent. We were using his car as we had put some plastic covering down in the boot, and then some straw to make it nice and comfortable for the little piggies. On arrival we found a well-organized farm with some impressive animal-handling facilities, so we took careful note of the gates and stalls that were used to handle the cows and pigs. The three weaners were quickly selected and placed into their luxurious travel cabin, and after the exchange of money and some form filling, we were off back to the barn! I'd initially worried that they might get a little jumpy in the back, and start trying to explore the car, but in the event they settled down immediately and slept all the way.

We'd fenced off the orchard some months before, and I'd checked the fencing before we went to get the pigs, tightening it up in places and putting obstacles where I was concerned the pigs might squeeze under the fence. At the time I felt it was more than ready to hold the three little piggies we were bringing home, and to a certain extent I was right about this for once. We released the little ones into the orchard, carefully picking them up and then quickly putting down the squirmy little monsters into their new area. I don't remember them making much noise at this point, but perhaps I've just wiped my memory to avoid thinking about any squealing. We threw some food into the pen and then decided to watch them for a bit. They were really cute, shuffling around and snuffling at everything they came near. They didn't seem afraid at all, and were quickly exploring all around the place.

Then I suddenly realized they could easily get out of the second gate, the one leading to our small field. I quickly grabbed up some mesh and ran towards the gate, but somehow they must have sensed the opportunity, and before I knew it, two of them were out in the field, running around like crazy. I really don't understand why they felt they needed to get out into that field, but as is often the way with animals, they'd done the one thing which would cause me the most consternation. Somehow I improvised some pig-herding skills and managed to persuade them back into the orchard area, and set about meshing the gate to prevent them getting out again. They seemed happy enough and carried on exploring their new home.

The area they lived in was half grassy and half covered in weeds, and I had decided that their role was to clear this area for me, given that pigs were legendary rotovators. I'd also read a number of books and articles which mentioned that you could keep pigs well on grass – that is, they would live off the grass and require little in the way of additional food. As was our way with the other animals we still wanted them to come to us when we needed them to, so we fed them once a day, but not a full ration, so as to encourage them to eat the grass. This they duly did, and while they didn't clear the whole area in quite the way I'd originally imagined – there were only three of them, after all – they still did a pretty good job.

Alex was both delighted and disturbed by her birthday present – delighted because they were such delightful creatures, and we finally had pigs, which she had been interested in for quite some time, but disturbed because they were already marked out for death: she felt it wasn't the best of presents as she wasn't getting to keep them. I doggedly argued that it would be great, as we'd get to look after them for about six months, and then she'd be getting all that nice pork as a grand finale to the present. But even after all this persuading, she didn't seem to think the situation was any better. Really, there is no pleasing some people.

As you might recall, the OSB was described as the perfect beginner's pig, but I'm not sure I'd agree. After that first incident they certainly didn't try too hard to escape, unlike, say, the legendary Tamworth. They were relatively biddable, and would follow me around the orchard area with or without food, though I guess they were always hoping for it! However, they had one trait which I did not appreciate – they liked to bite. They were particularly fond of biting my wellies, and while normally it was harmless, they occasionally managed to get my toe or ankle as well.

This was not their worst bite, though. Sue really loved the pigs and used to like throwing feed to them and watching them frolic in the sun. Then one morning she decided to feed them by hand, and unfortunately their bitey nature came to the fore and one of them sank its teeth into her hand! There was a lot of blood, but fortunately she was fine, and having learnt her lesson never tried to hand feed them again.

I think in our eyes this episode also took a little of the shine off the OSB. In addition, as they became bigger they became more rumbustious, and

I suspect that if we'd kept them to full size they'd have been difficult to control. It may be that our three were particularly excitable, or it may be the breed, but I didn't feel that I'd recommend them to beginners.

The Weaners Take their Holiday Trip

We'd had them for seven months when we decided to take them on holiday. Well, I decided, as Alex still refused to be involved in the death process. I arranged a slot with the nice people at the abattoir, and prepared for the day. Actually there was no preparation, because I'd decided they would go as they were and I'd tag them on the day to minimize the chance of them losing the tags. The morning of transport arrived, and Gordon, once again acting as animal chauffeur, backed up the trailer to the pig area. I put a couple of hurdles to either side of the ramp, then opened the gate and shaking a bucket of feed, walked up the ramp into the trailer. The pigs had a little trouble with the ramp but were eager to get to the feed, and were not fazed by the trailer; once I'd emptied the bucket in the back, they happily snuffled away at the feed. Gordon put a hurdle across the back of the trailer and I grabbed the tagging implement.

Tagging pigs is much like tagging sheep, although the forces of officialdom make it trickier by insisting on a new and different number for pigs. So now we had a CPI number to identify ourselves, a CPH number for our holding, a flock number for the sheep and goats, and a separate herd mark for the pigs. So we had sent off for a new set of tags, and all I had to do now was get them into the pigs' ears. I was rather worried that they would respond badly as their ears are much bigger than those of sheep, and seem more fleshy. Amazingly, however, the first tag went in without the pig reacting at all – it just carried on eating without a pause. The tag for the second pig was a little more trouble as it kept moving its head, though more in irritation at my handling its ear than anything else, but after a minute or two of trying I managed to get it in – and the third took only seconds.

So we were all ready, and the operation had taken less than half an hour and we were on our way to the abattoir. Clearly I was starting to get good at this. . . This holiday trip occurred just before Christmas, and we picked up our lovely consignment of pork at the beginning of the New Year.

As we'd been feeding the weaners mainly on grass, and they are a slow-growing breed, we didn't get the amount of meat that the books had suggested was likely, which I have to admit to being a little upset about; nevertheless we still got a huge amount, and were eating pork regularly for six months or so after that. It was good meat, though I suspect much of that was due to the fact that we knew where and how it had lived, as much as anything else.

Meat Yields and Animals

To me one of the most exciting parts of the 'holiday process' was selecting the cuts of meat. Laverstoke would provide a spreadsheet which would allow me to define what each cut should be, for example if the shoulder was to be whole on the bone, boned and rolled, diced, minced and so on. This not only introduced me to where the cuts came from (I knew about shoulders, but chumps?), but also gave me an idea of what we could get from each animal. I tried to find some diagrams on the internet, but most were too basic. Still, with first the lambs and now the pigs I had learnt a lot more about where exactly my meat was coming from. I have to admit I'd be thinking of how I was going to cook each cut of meat while filling out the form – and this was while the pigs were still snuffling around the orchard. Fundamentally I had never struggled with the fact that they were going to become food, and this was one obvious aspect of it. With pigs, however, selecting the cuts was made a little trickier as I hadn't a clue how big they would be!

When we'd had the lambs done I went for some fairly basic whole and 'on the bone' joints, as well as a few steaks and suchlike. The problem was the joints were fairly big, enough for four to six people, so I resolved to learn from this mistake with the pigs. When a lamb is taken for slaughter, one can expect around 40 per cent of its live weight back in meat; with pigs the yield is around 60 per cent, which is one of the reasons they are such a popular farming animal! Our OSBs were still relatively small in size (at least they seemed to be) when we took them on holiday, so I made the rather rash assumption that they'd be similar in size, if perhaps slightly bigger than the lambs, and based my cutting list on that. I was sadly rather wrong, and the leg joints especially were huge – each was enough for at least ten people! Fortunately I had asked for quite a few chops and steaks, but we still had some challenging joints. ... I think we put on quite a bit of weight over the following months as we'd try and tackle the smaller of the joints (even we would not try to split a ten-person joint between the two of us – between four, perhaps...).

All in all the pig-rearing experience had been very rewarding, and I felt that we'd learnt quite a bit, but most of all we'd been able to take care of them with little in the way of dramas. In fact I was ready to consider getting some more! In parallel with all this porcine fun we were also being broken in by the other animals, notably the cows...

THE COWS WIDEN OUR EXPERIENCE

When we'd let them roam more widely I'd originally been a little concerned that the cows might have some issues with the goats, but they just seemed to ignore each other most of the time. The only time they'd really interact was when I was feeding them, and that was to compete over the feed by all thrusting their heads towards the saucepan, a contest that Wrath would usually win, with Belinda the other occasional winner. However, when I spread the feed out across the goat tables for them they seemed happy to get on with eating it with no problems at all.

The Cows Go Walkabout in the Forest

The cows which had been in the field when we bought the place had clearly never really tested the fencing, given the state it had been in, and for some reason we thought it was unlikely our two 'little' ones would. Once again we were to be proved incorrect in our assumptions. I was at work one day when Alex, who was working from home, sent me this email:

> Email from Alex 8/1/8 17:15:
>
> went out about an hour ago to feed animals and cows were nowhere to be seen. torch just about to run out so coming back to get Land Rover to look for them when i heard a plaintive moo ... they too it appears have found the delights of the forest without the ability to get back. all 5 goats were there too...
>
> tried to lead them back but not having any of it so have cut fence – sorry – and top part of barbed wire on fence behind. still couldn't persuade them back but just don't want them to go the way of the sheep or to end up on the road. going to give them a bit longer – any bright ideas?
>
> must do this again sometime.
>
> a.

It was just after five in the afternoon so I immediately packed up and headed home. Now given my two-hour commute I wasn't home until just after 19:00, and as this was the middle of winter it was completely dark. There was also a light drizzle which made it even more difficult to see, as well as making the night wet and miserable. I quickly changed into my animal clothes (good clothes ruined by contact with the animals through either the dye from sprays, other stains, or rips from barbed wire...) and headed out armed only with a saucepan of goat mix and a recharged torch. Alex went back to where she'd left the Land Rover as that was near the big hole in the fence she'd enlarged when trying to get them back through. I climbed over the fence and headed into the woods...

As I made my way through the trees I watched the torch beam slowly fade and wink out, leaving me in almost pitch-black darkness. The rain had become heavier, and I had to try very hard to stop thinking about the Blair Witch Project. Even worse, it suddenly came home to me that I was looking for a black cow with a brown calf in the middle of a black forest. This seemed like a recipe for disaster, but I screwed up some courage and fought my way through the branches and twigs that were whipping me across the face. I couldn't hear anything other than the rain and my own passage through the trees, but it was even worse if I stopped because the darkness was complete, and the rain and wind made it sound as if there were things creeping around me; I had to try and rein in my imagination before I completely freaked out. A couple of times I nearly jumped out of my skin because I thought I'd bumped into the cows, *or something else* – but it was only a tree or a particularly woody bush.

I started calling out their names, partly just to hear the sound of my own voice and partly on the assumption (hope!) that they would recognize me – and I was soon rewarded with a lowing from Wrath which helped guide me towards them. I kept calling, and she kept lowing, and then I realized she was moving towards me, so I changed my tack to try and get her to go towards where I knew the Land Rover to be. This was relatively tricky as the wood is not flat, and as well as being on a general slope, it also has some undulations which are challenging to navigate even under normal circumstances, and impossible in such darkness. After stumbling around and suffering further tangles with trees, I sensibly went the long way round to try and keep to the flattest areas; it also helped that this was where the main footpath was. After a few minutes of this I spotted the Land Rover lights, which was a real relief – but also realized that Wrath had stopped.

I walked back into the woods, calling again, and once again she answered me and I could hear her moving through the undergrowth, heading towards me. I backed away, still calling, and soon had her coming into the light from the Land Rover, the first time I'd seen her. She carried on coming towards me and I eventually got her through the fence and back into our field, where I gave her a few large handfuls of goat mix as a reward. Alex was ready to close off the fence, at the very least by standing in the way, but Avarice was still on the other side.

I walked back towards Avarice but Wrath started following me, and since on no account did I want her back in the woods, I changed direction and managed to get her further away from the fence, dropping the goat mix on the ground to distract her. I then climbed over another part of the fence and carefully walked around behind Avarice, making sure she heard me and didn't spook, to encourage her back through the gap – and she slowly edged through the fence and back to her mother. Alex then quickly moved the fence back over and held it so the cows couldn't change their minds. Blessed relief, they were both safe and back home!

In the light of the Land Rover's main beam Alex and I pinned the fence back together as best we could, and she showed me where they'd pushed through. We put up a couple of pallets Alex had brought with her against the weakest parts, as a 'stop gap' until we could fix it more permanently at the weekend. Once again our fences had been tested and found wanting, and once again we resolved to do better next time!

One great positive about the incident was that it made me feel much better about keeping the animals, not because I'd been the hero of the hour (but really, that wood was dark and scary, and the twigs were sharp and scratchy), but because they'd come to me when I called, and I'd been able to lead them to safety. They might not trust me enough to allow me to hug them, or even to pour Spot On along their backs (Spot On helps to prevent flies, fleas and lice from bothering them), but they would follow me home.

This positive thought was very important as I was going through one of my regular bouts of worry about keeping the animals. There was always so much to do, and I didn't feel we knew anything like enough – my worst fear was hurting them because of our ignorance. But if they were starting to trust me then I couldn't be doing too bad a job, and with new resolve I looked forward to furthering our plans...

Stray from the Path at your Peril

As mentioned earlier, we have a footpath which crosses our field. It originally had a stile at each end, but I replaced these with kissing gates when I realized the stiles were rotten and about to fall apart. The problem with gates is that people need to close them, and even though we have put up large signs, a substantial number of people still seem to be completely incapable of closing a gate. This I find deeply irritating, especially when it results in one of the goats getting out.

The other problem with the footpath is that people seem incapable of following it: either they can't seem to find it, or they do find it and then wander off it. Through much of the year the footpath is very clearly marked due to the daily traffic of people walking on it, yet I have on several occasions had to explain to people where it is, trying very hard not to be patronising in the process... Worse than these people are those who think they have the right to roam all over the property, wandering over to see the pigs, or look at the goats (some people even feed them, which seems a bit rude and slightly crazy to me), or disturb the cows as they move in for a closer look.

93

Fortunately the cows occasionally help us gain some amusement from the footpath with the way they respond to people, and vice versa. One Sunday afternoon I was working around the fields (aka fixing fencing) when I heard Wrath mooing loudly, but heading away from me. I decided to see what was up, only to be greeted by a most unusual sight. There was a family of four, two parents and a boy and girl both around the age of ten, walking across the field with Wrath about ten feet behind them, mooing all the way. Now Wrath doesn't really moo in the way that children's audio books and television programmes might indicate – it is more of a roar, with only the slightest 'm' sound preceding it. It can be rather scary, and this poor family were clearly rather scared, but (certainly the parents) trying their best not to show it – I think they had decided that if they showed fear she might attack.

They managed to get to the other end of the path in the field and head down into the woods, followed by one last aggressive 'moo'. On their way down they passed another couple who were just heading into and across the field in the opposite direction, and they, too, were trailed by Wrath, who again mooed all the way! I think she carried on like this for an hour or so, and I was quite worried that we might get some complaints. Weirdly I have never seen her do this again, so I have no idea what triggered it – though there are times I wish she'd repeat her performance, particularly with those who wander.

Wrath is polled, and having no horns does not look too scary. She is still quite big compared to a sheep or goat, but physically her demeanour is not particularly intimidating – that is, until she moos. Avarice does have horns, however: they are about a foot long and stick out and forward from her head in a way which can only be described as threatening – just like the proper rounded ones you see on bulls in the Spanish bullfights/cartoons.

Many people have the view that only boy animals have horns, perhaps due to Walt Disney productions and children's television, which has meant that on occasion Avarice has been mistaken for a bull. She doesn't help the impression when she paws at the ground and lowers her head to bring her horns into position, but she seems only to do this when I try to stroke her. But one day we were inside the house when Alex called me over to the window to watch something funny unfold.

Right in the middle of the footpath was a couple arguing, clearly quite heatedly, but also quietly. Between them and their intended destination stood Wrath and Avarice, not really doing much, just standing around, chewing the cud, perhaps occasionally grabbing a chunk of grass, and mostly ignoring the arguing couple. It soon became clear that the woman did not want to go on, and this didn't change when Wrath moved off the path a little way, so we suspected they thought Avarice was a bull, and she didn't want to go anywhere near her (sorry, ...him). The woman pointed at her/him and shook her head vehemently several times, a scenario that didn't change for several minutes, until Avarice also wandered off the path to be with her mother.

The man obviously decided that enough was enough and strode determinedly past the two cows. The woman, now abandoned by her male paramour and stranded in the middle of the field, dithered for a bit and then tentatively walked around the cows as far as she could, going all the way to the fence to get maximum distance, before running to her partner and the gate.

I know it was cruel to laugh, but we both found this hilarious. As a result of this episode I was expecting a complaint from the council about keeping a bull in the field, but I've never heard anything! On a more serious note, I think people should always be careful of large animals such as cows and horses when they're walking near them. They are unlikely to attack anyone deliberately (unless their young are threatened, which everyone out walking should know), but if they are startled, say by a dog barking or the flash of a camera, then suddenly a half-ton of frightened animal could be running towards you, and that is definitely not good.

'Bulling'

Cows have a lunar fertility cycle similar to humans, and the gestation period for their calves is also nine months (leading the lady who sold them to us to muse that this might be the reason a woman is sometimes called a 'cow'...). They can show their readiness for a kiss and a cuddle in a number of ways, one of which is to mount another cow. However, a cow will not mount another which is more senior in the herd, which means that the cow at the bottom of the pecking order will not show her availability in such a manner.

Another sign is for them to call for their 'lover' by mooing loudly at about eleven o'clock at night. The first time this happened it woke us up abruptly (having only just drifted off) and we wondered what could possibly be the cause. Then we remembered that the nice lady we had bought them from had mentioned that the cows might 'come bulling' in such a manner, clearly from being so rudely woken up herself. Fortunately Wrath rarely came bulling in this manner, as I fear our neighbours might have had something to say if she did it too often!

Despite these two incidents the cows mostly seemed to ignore the footpath, and I think the fairly constant stream of walkers, especially on sunny days, helped to acclimatize them to people; certainly over the years they have become less and less bothered by both walkers and our guests. When we first had them we couldn't take more than one additional person out to see them at a time, because if we got within thirty feet of them they'd be off, sometimes going so far as to run off down to the bottom of the field. So the most that many of our friends saw of the cows was their two rapidly shrinking rear ends as they headed off into the distance – though I suspect most of them were happy enough with that!

The Cows Test the Fencing (Again)

After that first time in the woods the cows demonstrated to us on several occasions that they would continue to test our fencing and take advantage of any holes. One of these occasions was just after we'd had some sheep staying with us as before we sold our grass to a farmer, who took hay from the fields in the summer and grazed his sheep for a few weeks in the winter.

The sheep were brought over by a shepherdess, whose first task was to set up the fencing all round the area to be grazed, which effectively cut the cow and goat field in half – and which they were not happy about. This was to ensure the sheep didn't get lost, and didn't mix with any of our animals (though worms and the like can be passed through faeces, so the electric fence certainly wasn't protection against everything). However, it was nice having sheep on the field; they seemed very relaxed, usually just chewing cud and watching us as we went about our business. They studiously ignored the goats and cows, and were disdained in return.

One Sunday I was out tending to our recently planted hedge, and as part of the preparations I was trying to mulch it with some muck and small stones I'd collected (this was one of my early mulch attempts – it didn't do much to stop the weeds, but I think helped the growth a bit!). I was carrying the stones in a bag, and for some reason shook them out into a bucket. Suddenly every sheep in the field was running towards me – one instant they were placidly standing around, the next in full flight, directly at me! 'Stampede' was a concept that ran through my head, and I wondered what would happen when they got to the now rather thin-seeming electric wires which were the only thing between me and almost certain oblivion beneath their galloping hooves.

Our main field was split up by a very basic fence which we'd patched up a little, but had opened at the bottom to allow the sheep free access – but in their eagerness to get to me they chose not to go the long way round, but ran over the fence, in several places knocking it down completely. One or two fell over in the process but quickly righted themselves. I hadn't moved at all while this was

TOP: *Belinda, recently sheared and standing on our goat patio.*

BOTTOM LEFT: *Bill being hand-fed strawberries, while Ted looks on a little sceptically.*

BOTTOM RIGHT: *Ted has realized there's food to be had and wants to get as close as possible!*

TOP: *The first lambs to be born at Shaw Barn, looked after by their very protective Mule mother.*

BOTTOM: *Horny and her triplets.*

TOP LEFT: *Bertie (left) and Boris, with their fleeces grown out and a little shaggy.*

TOP RIGHT: *Me, concentrating on the tricky procedure of castrating a day-old lamb.*

BOTTOM: *Four chicks just days after hatching, enjoying some chick crumbs out of a ramekin.*

TOP LEFT: *Geese and ducks hanging around the slightly battered chicken ark where they'd all been hatched.*

TOP RIGHT: *Our first gosling sitting prettily.*

MIDDLE: *One of our broody bantams looking after the two eBay ducklings.*

BOTTOM LEFT: *Little lost sheep – Mouton in her first few days with us.*

BOTTOM RIGHT: *One of our Suffolks, laughing at the absurdity of it all.*

Top Left: Muga, a Soay ram and the daddy of all our lambs.

Top Right: Mouton with our first Soay lamb.

Bottom: Whiteface, escorted by our three original Soays: Mouton to the right, Lafite to the left, and Muga leading.

TOP LEFT: *Wrath and a very young Avarice just after they'd arrived.*

TOP RIGHT: *Wrath sometimes gets close enough to sniff at my hand, before backing away again.*

BOTTOM: *Wrath and Avarice looking rather suspiciously at me, with Bertie in the background looking on in hope of some food.*

Top left: Our first piggies – the Oxford Sandy and Blacks.

Top right: Sir Humphrey (left) and Bernard.

Bottom: Alex talking to Percy.

TOP LEFT: *Sir Humphrey lazing in his hut in the winter sunshine.*

TOP RIGHT: *Gaffer (top) and Hacker. If you look closely you can see a clear triangle of white in the shadow just above Gaffer's front leg from her early injury.*

MIDDLE: *Sir Humphrey enjoying a wallow during the peak of summer.*

BOTTOM: *Snowball allowing her litter to suckle.*

happening, other than to turn round – and then just as suddenly they all stopped, as if they had hit a wall, about thirty feet away, and still some twenty feet away from the fence. It was as if they suddenly realized they didn't know me and would therefore, on second thoughts, probably not want to have dinner with me (as I assume they thought the noise was food in the form of sheep nuts); but the amusing thing was they then tried to be nonchalant about it, slowly drifting away and avoiding looking at me until they were all back where they'd started, and I was back working on the hedge as if nothing had happened.

After a few weeks of having the sheep as guests the shepherdess came to take them away, which involved her loading them all up into a smart double-decker trailer and taking them to a different field; she then returned to take up the electric fence and wind it all back in again. It was during this process that the cows demonstrated their sneakiness and speed. We'd fixed up the sheep-crushed fence so there was only one gap between the two fields, right at the bottom, and that was covered by a gate. While the shepherdess was gathering up the electric fence she'd left this gate open, having checked that the cows and goats were all at the top of the field. But a few minutes later that had changed. Gordon and I were working on the barn, trying to sort out the owl box I'd bought to encourage a barn owl to come and live with us – though on the outside of the barn – in an effort to control the rodent population. I heard a soft moo and looked round to see Wrath standing only a few feet away, looking at us rather curiously.

This was a shock, as the nearest she should have been was well over a hundred feet away – I had no idea how she could have got there and was a bit stunned. In retrospect it was clear that she'd run down to the bottom of the field, through the gate, and then all the way back up again, possibly just to find out what the box was on the side of the barn, or in the hope of some food. The problem we had was that the field she was standing in had no fence on one side, its boundary was the driveway, and it was also open to Rob's fields and, much more worryingly, the main road!

Fortunately the head builder from next door (they were in the middle of a major renovation at the time) had just come round to answer one of our many questions, and realizing what had happened, moved to head down the path and around the cows (Avarice was standing behind her mother). I started to move towards them, but very slowly, and then stopped when I realized that I could only push them towards the road. Then I noticed that the shepherdess, realizing what had happened, was racing towards the gateway to the road in her truck, and would soon have it covered.

Suddenly Wrath decided she didn't like all this attention, turned round and sprinted down the field, first towards our builder friend, before veering back towards the gate she'd come out of, but then swinging round towards the gate to the road – which fortunately was now covered. Avarice and I started running after her (with me last in the race), but fortunately Wrath chose to head back to home pastures. I arrived at the field gate at the same time as the shepherdess, who understandably was very apologetic. We'd been very lucky that Wrath hadn't headed straight for the road, but fortunately disaster was averted and we'd learnt another valuable lesson – that our cows could certainly shift when they wanted to!

We also realized that we needed more fencing and another gate in the field boundary near the driveway to stop a repeat of the problem, and also so we could use that field for some more animals. After attempting to put up our own, and an experiment with some more electric fencing, we thought it best to pay someone to do the job for us. They put up a lovely post and rail fence which made it look like a proper boundary. We also had a full gateway area put in so that we could move a trailer in and out without too many challenges, and so vehicles could pass each other easily on the driveway.

With the extra confidence from the new fence, and a desire to get a sustainable source of meat, we looked again at sheep!

EXPANDING THE SHEEP FLOCK

At the same time as we added the OSBs to our holding, we also started building up our flock of sheep, continuing a trend of adding animals in waves as we became more confident. The experiment with fattening up the lambs had given us the confidence we needed in handling sheep, and we knew that Lafite and Mouton really needed some friends as just two of them really didn't count as a flock – at most we could describe them as perhaps a 'flockette'. So we talked with our farmer friend and fount of all knowledge, John, about getting some more.

We Purchase A Small Flock

After some discussion his advice was that we take some ewes with lambs at foot (that is, still suckling). This would mean we'd have some lambs to fatten up in the first year, and then, once we'd decided how we would breed from them – in the sense of whether we would use a ram or AI – we'd have lambs each year going forward. We liked this idea, and therefore only had to decide how many couples we wanted. Our twisted logic escapes me now, but we settled on needing four ewes, expecting to produce six to eight lambs a year. Gordon asked if he could buy a ewe too, and we would then look after it on the same basis as the previous lambs and OSBs (shared costs and so on). We agreed, so we settled on getting five ewes and were expecting around ten lambs with them.

On the day John rounded up the sheep from his flock, five ewes and their eight lambs seemed to naturally separate themselves out for us, and so those were the ones he brought to us. Initially I was disappointed that it was only eight lambs, but then realized that as we now had fifteen sheep (including the two Soays) they would probably be enough of a handful! The new flock was duly released into our small field, to keep them separate from our existing animals in case of foot rot or some other issue. We had finally learnt a lesson! This field should have been cut for hay, but hadn't – which leads to a small digression into hay.

Managing a Hayfield

At this point we still weren't using all our land. As mentioned before, we had a verbal agreement with a local farmer (not John) that he would use the fields for hay, and then for grazing his sheep during the winter for a rather small sum of money. The idea was that it would keep our grass mown and the weeds mostly under control, and a small sum of money is better than nothing! It had worked out fine the year before and we were expecting much the same again. Unfortunately our poor fencing had meant that the cows had got into the hayfields and wandered about in them for a couple of weeks before we could sort it out, and it appears the farmer had seen them. He'd cut our neighbours' fields but then not ours, and after a couple of weeks I started to get worried.

I decided to phone him (yes, shocking indeed), whereupon he told me that 'you can't make hay from sh*t', and went on to explain that because the cows had been in there it would have ruined the grass for hay. This was complete news to me, and I was a bit annoyed that he hadn't told me beforehand or perhaps when he first saw the cows in it; a simple injunction to keep the hayfields animal free would have been useful, or indeed any form of communication. He did grudgingly offer to top the fields later in the summer to ensure they didn't go totally wild (which he never did), but this was the beginning of the end of our arrangement with him.

We also later discovered that he was paying us about a tenth of what we really should have been getting for the hay and fields, which was disappointing, especially as it validated the stereotype of country folk taking advantage of naive city folk. This was finally the end of our agreement with him!

The Ewes Settle In

So the new sheep were in the 'was-to-be-hay' field, and soon went about investigating their new home. One of the things I worry about is whether new animals will find the water. It's rather like moving to a new office, or joining a new company: it's important to find out where the toilet is, but quite often no one mentions it and one ends up wandering the corridors, potentially in some distress, searching for the facilities. Every book on every animal is insistent that they must have clean water available at all time, and we've been pretty good at ensuring a fresh running water supply. Sometimes it's running because it's leaking, but at least it's there! Setting up water drinkers is probably the next most time-consuming job after fencing. The drinkers themselves get knocked about by the animals, and occasionally humans, so the pipe connections loosen and leak, or they get pushed to an angle and so overflow. The pipes can also cause problems if they aren't buried, with animals kicking them and breaking connections, a particular favourite of the pigs. The worst is when things thaw out after a freeze, when we tend to get leaks all over the place!

With our new sheep there was no obvious way to lead them to the water, as they weren't yet used to coming to us for food. I thought it might be best to herd them into the corner where the water was so they'd see it and know where it was in the future. Once again I became a human sheepdog, running up and down the field until I'd eventually managed to chase a few of the new ewes into the corner with the water. I thought the ewes needed to see it as they would be more responsible than the lambs. Looking back I suspect they were so stressed at being chased around the field that looking at water was the least of their worries – but it made me feel better. I don't do this any more as I've realized that when they explore a new field they test every inch of the fence and will therefore always find the water (as we always put it next to a fence to keep the piping exposed to the minimum).

All the agricultural tales and fables would seem to indicate that sheep are stupid; it's one of the few country facts all city folk know, and we were still prejudiced when our new sheep arrived. Our original two ewes, Lafite and Mouton, were Soays, and known for having a bit more character and intellect, and so were exempt from being considered stupid. The new sheep, by contrast, were 'mules', a commercial breed which clearly would have had all sense bred out of it in favour of meat yield – and so we were quite surprised when some of them seemed to be rather sharp.

As was becoming standard with our animals, we gave them additional grain-based feed each morning to get them used to us, even if they were in a field full of lush grass. We were still feeding them goat mix at this point, and they all really loved it; it's much tastier than grass, and contains much more energy. However, it was around this time that I was reading one of our books on keeping sheep, and learned that we should in fact definitely *not* be feeding the sheep with goat mix, as it contained copper.

We Revise our Feed

Up to this point we'd always been a bit laissez-faire about the feed we gave the animals, as everything loved goat mix. We'd therefore simply bought goat mix and fed it to everything, despite it being one of the more expensive feeds, since we felt it was important to keep them all happy. We learned that the problem was that sheep are very sensitive to copper, and if they eat too much it can build up in their livers and kill them (more rarely, if they don't have enough it can cause complications when they are lambing, but one problem at a time!). The copper in the goat mix was in one of the grains they used to make it, and so we really had to stop feeding our sheep with it.

We replaced it with ruminant mix, a similar sort of grain mix which the animals liked, though they didn't go mad over it as they did with goat mix – I suspect because it wasn't molassed in the same way and so wasn't so sweet! We also added sheep nuts as these would provide some variety and were chock full of energy and so should help to fatten up the lambs, and keep the ewes healthy. They were slightly cheaper, too, which was becoming more of a consideration as the number of animals increased.

Managing the New Flock

After the false start with goat mix we settled into a good rhythm with our new flock. I'd walk out in the morning, pour out the feed in three or four piles, and the ewes would dive on to them. One part of feeding used to drive me mad (and still does), and that is, as I put down a pile they'd dive on it, then I'd move along and put down another pile, and they'd all run to it, leaving the first pile free … and this would go on for three or four piles. It was as if they considered there was additional value to the newest feed being delivered, and it meant that while I was pouring out the feed I would be mobbed, which could be frustrating, to say the least.

At first only three of the ewes would come to feed, with their lambs tagging along, then after a couple of weeks the fourth ewe joined the party, if at first a little tentatively; she was distinctive because she had horns, one of which was slightly twisted, and so she was dubbed Horny. The last ewe with her two lambs would not come within thirty feet of me. She had a distinctive white stripe down the middle of her face, and I named her White Face (my imagination was clearly on holiday during this period). She was to become my favourite of all the sheep.

White Face was clearly not stupid: she had been dragged to a new home and now a stranger was trying to tempt her with treats, but she obviously considered that clearly this was a ploy and she was not going to let herself get sucked into it. It took a few weeks for her even to sniff the food, and although she clearly liked it, she still didn't rush in the next day to fight for it, but would stand a short distance away with her two lambs and stare suspiciously at me, occasionally bleating, until I'd headed away again.

This was starting to be an issue, as I really needed to catch them and spray them with Crovect to protect them from fly strike. We'd bought a set of sheep hurdles (the small fences you might have seen on the television programme *One Man and his Dog*), and I set them up in a box-shaped pen around the area I'd been feeding them in, with the existing fence on one side to make the area big enough to keep them all in. The four friendly ewes were a little disconcerted, but still carried on coming in, but this put White Face off even more and she retreated back a further twenty feet or so. Still, after another week she was coming to the entrance to the box, so I felt it was time to try and trap them and spray them.

My plan was to feed them, and then walk round and herd White Face and progeny into the box and then close it up. Amazingly it worked first time, and I managed to catch all of them in one quick movement of the last hurdle, which closed to form the gate. Alex, who'd held back during this process to avoid spooking them, then ran up with the Crovect, and we proceeded to spray them. We made a rookie error at this point in that we still let them have the full area of the box, which made controlling them quite hard as they ran around within it. The basic technique was to spray them as they went past, and in effect stir them around until they were all covered, though not entirely evenly!

Over the next few times I learnt to close the box down into a tight area where they couldn't move, then we'd pull them out one at a time to spray them, cut their hooves, or whatever it was we'd determined we needed to do. Satisfied they were Crovected now, we released them. However, White Face was clearly not at all happy, and she stayed even further away for the next few weeks.

I think she was most worried about her lambs, because once they were big enough and started being more independent she started to edge closer – and then one day it was as if a switch had flipped, and not only was she coming for food, she was at the front of the flock, trying to push her head into the feed bucket! From that day on she was the most forward of our flock, always pushing to the front, and demonstrating she was the flock queen. Each morning she would greet me first, often with a friendly bleat, and sometimes climbing up on the fence to get closer to the food. She'd also run up to me even at non-feeding times, in search of food I would guess, and while she'd quickly realize if I had nothing, she might hang around for a while to see if I was doing anything interesting.

For a 'stupid' animal she'd clearly worked things out and decided two things: first, that she'd landed on her feet in this new place, and second, that I was a soft touch and good for food! We were talking with our shearer about sheep and their rumoured lack of wits on one of his visits, and he said to us that all stock is wise when it comes to food, and I really can't argue with that.

We Purchase a Ram

Now that we had a set of adult ewes – five mules and two Soays – we decided that we needed to get a ram to ensure a supply of lambs, and therefore of meat! After a little research and some deep thinking, we determined that we'd get a Soay ram for our flock.

This was for a number of reasons, starting with the fact that we liked the Soay breed. As may have become clear in the course of my writing so far, we like animals with character, and we have been blessed with quite a few of them! While White Face had some tricks, some of the other mules were far more placid, and we wanted a little more of the character which the Soays definitely had. In addition the Soay is a rare breed, and while not as close to extinction as it once was, it is always good to keep these breeds going if you have no particular requirement for meat yield. I'm not zealous about rare breeds (otherwise we would never have had the mules), but I do think there is some logic in keeping a wider gene pool available. There are certainly some rare breeds where the gene pool has become tiny, and there may be arguments for breeding in some other lines to give them some additional strength, but that's a long and complicated discussion that we really don't have time for here.

In addition we certainly felt that it wasn't sensible to get a ram larger than the Soays, as that might cause them problems come lambing; nor did we want two rams. We thought that a smaller ram covering the mules would make lambing much easier as we'd heard that many of the problems with lambing come from the size of the lambs. Finally, having heard that Soay meat was stronger and tastier than normal lamb, we thought that this, crossed with the mules, might give us lambs with a good flavour, but which weren't too small. Our variety of reasons for selecting a Soay ram may therefore have been a bit woolly, but that didn't make them bad.

We were set on our decision and consulted the Soay Sheep Society's latest booklet to see what was available. We'd joined the society early on thinking it might be useful to learn more about the breed, and we had been considering registering our ewes and breeding it as an official flock. In the end we chose not to register our sheep but just to read the articles, which were often amusing or informative, if not both! A woman in Dorset was advertising Soays, including some rams, and she was the nearest to us of the various offers, so Alex contacted her and arranged for us to visit. We duly headed down in our trusty Land Rover, and by some miracle found the place, which was tucked off a lane and behind a farm. It was a little smallholding with a few fields that supported a flock of Soays – there must have been about thirty of them – and some geese and chickens.

We met her main ram, and a fine-looking specimen he was. She had two younger adult rams for sale, one of which was registered with the Soay society. They were typical Soays, brown with the pale patches around their bottoms, but with bigger, thicker horns than we were used to in our females, and we made sure they were both intact (this is an important check, as people have bought castrated males thinking they were rams...). She also showed us some younger rams, including a lovely little piebald fellow, white all over with brown and black patches. Soays do have some variety, and piebald is one of the rarer types. We were both taken with him, but needed to think about it as he was very young and would not be able to serve our sheep that year. We left her with a promise that we'd get back to her the following day with our decision, and headed home.

On the way home we debated the pros and cons, and agreed that while a shepherd with more commercial leanings would not want to miss out on a season and would therefore take one of the adult rams, we were in no actual hurry and we really liked the small piebald. We thought it best to sleep on the decision, and after a restful night we still agreed on the little one. I can't remember what we were doing that day, it was a Sunday and they are usually busy, but Alex wasn't able to call the woman in Dorset until the evening, whereupon she gave us some unhappy news. Her husband, who'd been out when we were there, had decided to get all the castrations done that day, and had castrated the piebald as well!

If we'd known of the risk I think we'd have taken him straightaway and brought him home in the back of the Land Rover, but we'd missed our chance, and now needed to make a decision on one of the other two. We settled on the unregistered ram, as our ewes were unregistered and we'd already made the decision not to have a registered flock. The next weekend we drove back down to Dorset and picked up the ram, which the woman had kindly registered for us as well, and passed on the cost to us. I am not sure why our communications went so awry with her, but anyway, we now had a ram!

We named him Muga after our favourite Rioja (a Rioja of the old style with rich raspberries and cream flavours, and which I can definitely recommend). Given we'd named the first two after Bordeaux wines, this nomenclature seemed to make sense to us. He was a feisty little ram, and we thought he'd do very well with our flock. We duly released him into the small field with them; by this time Mouton and Lafite had also moved into the small field, though we weren't clear how! Nevertheless this was good news as all the sheep were now in one place. Muga quickly introduced himself to the flock, and after checking to ensure there were no other males, he assumed his rightful place as the leader (though of course White Face was still really running the show...). We'd heard tales of rams and how they can be aggressive, but he seemed pretty relaxed and we were soon comfortable with him, though we became perhaps a little overconfident.

Like all the sheep, Muga loved the ruminant mix and ewe nuts we fed them, and he would also be quite pushy in trying to get to it. On one occasion when I was walking the mix a bit further into the field than normal he became rather impatient and rammed me! He only backed up a short way, and I didn't see him charge until suddenly it felt like a rock had slammed into the side of my thigh. It was a real shock, and he started to back up for a second go, until I poured some food on the ground and then he immediately went for that and was fine. I was very careful after that and warned Alex. She, of course, considered I was fussing, and to this day he has never rammed her. She revealed her technique as having had severe words with him – rushing towards him, then going down to his level, holding his horns and shouting!

I, however, have taken another couple of hits since then, once when I was slow with the food again, and once when I was planting clover seeds in cowpats around the field and he decided I'd taunted him with the offer of food (it was a rustling packet of seeds)! On each occasion it was the shock more than anything which got to me; it didn't hurt that much, and although I ended up with some bruising it was nothing serious. He never seemed to bother the walkers in the fields, which was something I was concerned about after the first episode.

I think he realized they weren't food sources and would soon be gone. He'd also ram any creature which got in his way when food was about – sheep, goats or cows, they all got it in the side at some point or another. The cows especially could be quite flighty around him when he was being particularly aggressive.

Three More Soays

As we headed into autumn we were happy that the flock was bonding well, and were looking forward to our first crop of lambs. We had a call from the woman who'd sold us Mouton and Lafite to say that she had three more and were we interested? Well of course we were . . . again I'm not sure we put much thought into it, but at the very least we felt it would give the two Soay ewes a few more friends of their own size, and they were small, so wouldn't eat that much, either grass or feed, and we knew that the woman really wanted them to live and not go into the freezer – so off we went to add more to our flock. . .

We were getting a real handle on picking up animals, and it seemed to take almost no time at all to go to Somerset, pick them up (which did include another flying tackle or two to grab the flighty little things), and then drive them all the way back – all in all a five-hour round trip. The first two were twins, which we named Latour and Margaux (retaining the 'first growths' naming scheme), and which I've never been able to tell apart. They're very nervous and prefer to stay at the back of the flock at all times. The third we named Haut-Brion, thereby completing the set. She has a much darker face than the others, and is a little more inquisitive. As with all our animals, she was shy at first, but once she felt more comfortable she would join Mouton in coming closer to me to see if there was any extra feed, and she'd happily eat from my hand, if only in two-second bursts.

Sorting Out the Mule Lambs

In early December we were ready to take the mule lambs on 'holiday' as they'd grown into good-looking sheep and were no longer traipsing around after their mothers as they had when they'd first come to us. They were nearly adults now, and showing off their independence. Fortunately they were also addicted to the ruminant mix we fed them, so we knew that catching them in preparation for their holiday should not be very tricky. However, it was still challenging.

We fed the sheep every day in the same place, in an area surrounded on three sides by sheep hurdles, which meant we should in theory be able to close the fourth side, and 'hey presto' they'd be stuck in our little enclosure. The problem was that they tended to swirl into the enclosure, take just a quick lick of the feed, and then as the others forced their way in, some would eddy out – rarely would all of them be inside in one go. Obviously we only needed the lambs, but this was actually more challenging as it was the mothers who tended to stay in the area, eating the food, and blocking the lambs out... Still, we'd done it when Crovecting them, so we knew it was possible, we just needed a little luck and good timing.

I helped the luck by taking more food than usual and putting it in more piles, thereby keeping more places for them to stay. The tactic worked, and I managed to close the hurdle before any ran out. I then deployed a new technique to make controlling them a little easier. One of the problems we'd had when catching them in the past was that they ended up running around the pen, which put pressure on us while we were ministering to individuals, and meant that on occasion they would find a weak part of the hurdles and manage to push through and get out. To make matters worse, some of them were even sprightly enough to jump over the hurdle completely.

However, I had watched a couple of episodes of *One Man and his Dog*, and learned that if I took an extra hurdle into the pen I could herd them all into one corner and hold them there, and if they were all packed together, and the hurdles were sturdy enough, they couldn't get the leverage to push too much, let alone leap over. This was surprisingly easy to do – and I was suddenly, for the first time I could remember, completely in control of the sheep! I used my power wisely, quickly separating out the Soays and the mummy sheep, and leaving the eight lambs.

Crutching and Tagging

Our abattoir had changed their rules (or perhaps we'd not realized them earlier) and they now demanded that we 'crutch' the sheep before bringing them in. This involves trimming around their rear ends to get rid of any of the poo and other stuff which tends to clog there. I'm not sure why they insist on this, but it was now the policy, and I didn't want to be sent away for having lambs with dirty bottoms! I also needed to tag some of them as they'd lost their tags. This was my first time for crutching, and I was a little nervous. Crutching was back-breaking. I had to grab the lambs out of their enclosure and bring them out while reclosing the hurdles, bend them back into me in a sitting position, and then lean over to get to their bottoms to clip them, before returning them to the enclosure. It took me over an hour to do all of them, and I was truly exhausted by the end of it.

Then I had to tag them. We had brand new tags and plier device, and all I had to do was fit the tags into the pliers, hold a sheep's ear in between the tag's 'jaws' and apply pressure. Simple! Well, not really. The lambs weren't exactly cooperative and would flick out their ears just as I was closing the jaws. I also found that if I didn't get the angle very close to 90 degrees I'd end up bending the sharp bit of the tag and failing to get it in, or get halfway through the ear but not connect. After another thirty minutes or so, and around twelve broken tags, I'd finally finished the tagging, and all I needed to do was get them into the trailer and we could be off! Fortunately Gordon, once again acting as animal chauffeur, had backed the trailer up virtually all the way to the enclosure, and with a little food, and some gentle pushing, I quickly had them into the trailer and the back closed!

A short sigh of relief and we were off to the abattoir, only running about an hour later than planned. A quick call to the abattoir allowed me to relax a bit, as they said that although I'd lost my slot, I could still come along at the later time (I think I'd have cried or screamed in rage if they'd said I needed to come another day!). The off-loading was an absolute dream, and we were quickly back on our way home. Two weeks later we picked up all our vacuum-packed meat and started tucking into some delicious meat!

Thus at the end of the year we had what was turning out to be a reasonable flock, with ten ewes and a ram. We were looking forward to having our first lambs born in the spring, with the possibility of having up to twenty lambs (assuming they averaged twins), but the likelihood of it being closer to twelve or fourteen. Eleven sheep was clearly enough and we had no intention of getting any more.

Our neighbours, Rob and Claire, had a flock of sheep as well, which they'd bought off the farmer who'd been renting our land for hay, about eighteen months before. They were Suffolks, and there were six of them. Suffolks are much more solid than the mules we had, with a thicker and tighter fleece and a slightly chunkier rear end, which I imagine means they tend to yield more meat. They also had some English Saddleback pigs, as well as horses, chickens and three children, so they had quite a lot on their hands! That winter was bitter, and all the water pipes froze, even the ones in their barn, which meant Rob or Claire had to trek back and forth from their kitchen to get water to give to all the animals each morning and evening. After a month of this pain, and the general burden of the year before, they decided to pack it in, and put their pigs up for sale, and offered us their sheep.

The Flock Expands Again!

Let us take a brief pause here. Alex and I were still both commuting to London, a return journey that took four hours every day, as well as working jobs which often demanded late nights (or in Alex's case, sometimes all night) and weekend work. We already had three goats, two cows, a handful of chickens and eleven sheep, plus the dogs, cat and fish (just the three angel fish at this point).

We'd recently discussed that we were always busy, tired and running from place to place just to stand still. Our fencing was quixotic at best, and we'd also agreed that we wouldn't dive into things again, and would plan things out and be on top of everything. So of course we said yes to Rob and Claire's sheep, and could we have some of the pigs too? (More on the pigs a bit later.)

They delivered the six ewes to us the following weekend. Well, what actually happened was that the sheep got out on to our shared drive while Rob was tagging them, and we decided we might as well herd them into our small field while they were out and save ourselves some time! One of the ewes was called Lamby (and at one point had been called Thunder by Callum, one of Rob and Claire's sons); her mother had died when she was born, and so Claire had bottle fed her. She'd tried to mother her on to another ewe but it hadn't worked, so she'd fed her bottled milk several times a day for the first few months of the year. I suspect this contributed to their decision to give up sheep. Still, Lamby was a nice ewe and even responded to her name – though I do wonder what would have happened if she'd been a boy... Perhaps they'd have kept him as a ram or perhaps as a 'real pet' after castration.

Once the ewes were in the field Muga went straight over to them to welcome them to the flock in his own inimitable manner – and they very quickly integrated. I think once the flock gets to a certain size new sheep settle quickly. It also helped that these ewes had been well handled and so were not wary of human contact, as all our previous sheep had been.

Thus we started the year and entered our first ever lambing season with sixteen ewes and a ram, a few more than originally planned. Nevertheless we were quietly hopeful that we were ready for the challenges ahead...

MORE CHICKEN EPISODES

Of all our animals, the chickens required the least effort to look after. We had four hens left: Bella Bella, Cher, Luke and Beau, and while they didn't have a cockerel looking after them, they seemed content with their lot in life. They were producing two to three eggs a day, which were delicious, and we seemed to have found a working method for looking after them. During the summer and early autumn we let them out of their run in the morning and put them away at night, but in winter and early spring when the days were shorter we'd only let them out at the weekend to make sure they were never at risk of being taken by a fox at twilight and into the night – or at least we reduced the risk.

Then one day towards the end of summer we noticed that two of them, Luke and Beau, were not coming out to feed, and we were a little worried about them. But on opening up the ark we found them both sitting quite happily in the nest box, with an egg each – they were broody!

eBay Eggs

Bantams are known to make great broodies because they like sitting and will, most of the time, 'go the distance' – we'd read that other birds, and notably geese, can seem to get bored and wander away from their eggs. This presented us with an opportunity to increase our peep (my favourite of the collective nouns for chickens), which we were keen to do as we wanted to try again with a cockerel. We also felt that four wasn't really a big enough group, and wanted to get back to seven or eight birds. As we didn't have a cockerel obviously none of their eggs would be fertile, but if we could find some fertilized eggs then our broodies should be able to turn them into little chicks. Once more the internet proved a great source of information, and indeed eggs, because we bought our eggs on eBay!

It's not legal to sell live animals, or eggs for eating, on eBay (presumably due to health and safety regulations), but eggs for hatching are just fine, and there is an amazing array of them. The great thing about eggs is that if they've been fertilized they can stay in that ready state for several days – indeed long enough to collect half a dozen, put them in a special polystyrene box and send them halfway across the country! Deciding which eggs to buy, given all the choice, was difficult, but in the end we stayed with the Pekin bantams, and bid for, and won, six eggs from yellow and lavender Pekins. In no time at all we were told the eggs were in the post!

Only two days after the auction ended the eggs arrived, packed in a polystyrene box, then with bubble wrap around them and all placed in a jiffy bag, tightly wrapped in sellotape, and finally in a normal brown envelope – there was really no way these eggs were going to be damaged in the post! In my excitement I took them straight to the chickens and gently placed the eggs underneath them, removing the ones they were already sitting on. They didn't really appreciate me removing their eggs, indicating this by pecking at my hand, but quietened again when I put the new eggs under them. In the thrill of receiving the eggs I'd forgotten that the recommended practice is actually to leave the eggs for a little while, some say up to a day, before putting them under the hen, to allow the eggs to settle down after being shaken up in their journey. Hopefully next time I'd be a bit calmer and remember!

Bantam eggs take about twenty-eight days to hatch, sometimes a bit longer. Luke and Beau were diligent in sitting on the eggs. We checked them every day, usually giving them some extra feed by hand, and even giving them their own little water cup so they didn't need to head downstairs if they didn't want to. The first couple of weeks all seemed to be going fine – but then we noticed a little mite on one of the eggs.

The Killer Red Mite

The mite was a red mite. These are normally brown, but become quite red when they've recently fed – on chicken's blood. We'd never experienced them before and didn't worry too much about it; there was only one mite after all, and the book I checked in said they were more of a nuisance than anything else. But the next day there were a few more, and so we looked into what we needed to do in order to control them. We bought some powder, which was supposed to kill them, or at least discourage them, and sprinkled this liberally around the nestbox. I think this did limit their spread, but it didn't stop them all as they live in cracks in the wooden structure of the hen hutch, and then transfer to the chickens. Over the next few days we became more concerned as their numbers weren't really diminishing, and we were worried they'd prey on the chicks when they were born.

Then one morning when we went out to check on the broodies we found them still sitting on the eggs, but they were completely still – and we realized they were dead. It appeared that the stress of sitting on the eggs (they have to provide extra warmth to the eggs, and they can eat and drink less frequently), as well as the predations of the red mites, had finished them off. We were devastated: not only had we lost a batch of eggs, but we'd lost two of our remaining hens, both of which had been with us for quite a while.

111

We did more research on red mite, and several sources mentioned that they could kill chickens – which is one of the reasons I now tend to check multiple sources whenever I spot a potential problem with one of our animals. We learned that getting rid of red mite is very difficult and requires constant surveillance. Basic controls include the powder we had used, and a spray. Preventative measures included recommendations to creosote the chicken hut – although this was often described as an old practice, as recent recommendations were to avoid creosote completely to avoid its constituent toxins, which has made creosote rather hard to get. At this point we used just the powder and spray, and through judicious use of both seemed to wipe the red mite out in our ark. Nevertheless we always keep a keen eye out for it after this incident.

New Additions

This was a real setback for us and dented our confidence somewhat, and it was several weeks before the situation changed. Alex took the loss of the chickens quite hard, and observed how lonely and sad the remaining two looked, so I decided it was time to cheer her up and get some more. I scoured the web until I found someone nearby selling chickens, phoned them up to confirm details, and then jumped in the car to get them. Unfortunately the place was more than an hour's drive away, but once I got there I was amazed at the set-up in the lady's fairly large back garden: she had a complex of huts and meshed areas, with at least four separate blocks of fencing, with each block subdivided further to allow her to keep, for example, the cockerels separate from the hens. She had a great selection of chickens, big and bantam, and we soon agreed a selection of five or so, including a cockerel, and I headed home with a big smile on my face!

Alex was really pleased with the new chickens, and they quickly made friends with Cher and Bella Bella. We called the cockerel Boss Hog, keeping, belatedly, to our 'Dukes of Hazzard' naming scheme. I jokingly asked one of my friends if he wanted to buy some chickens, and he went with the joke and agreed to buy two, naming them Emma and Cassie. The agreed payment for boarding the two chickens was their eggs, which was clearly a good deal. Emma and Cassie were black and frizzly, a bit like Cher, and the other two, which I didn't name at this point, were both yellowy orange. So once again we had a nice-sized flock, with a very lucky cockerel!

A brief aside on naming: when we first started getting animals we named everything, except some of the smaller fish, almost obsessively. However, as our menagerie increased we realized they didn't really care what we called them, and in fact the majority never respond to a name, even if they might recognize it! So we tended to name only those creatures which either stood out for some physical reason, such as the cockerel, or which demonstrated some extra character, such as White Face (and that really isn't much of a name!). On occasion this did cause us some issues when trying to describe a particular sheep or pig, but generally we managed by pointing and trying to find some physical distinction. I did feel slightly guilty when an animal died and we could only refer to it as that dead yellow chicken (for example), but I guess this was just me being oversensitive. We also tried not to name an animal we planned to eat, especially after the Clarissa incident.

An Unfortunate Combination

We were looking after my sister-in-law's dogs for a month or so at this point, and I was trying to train one of them, Buster, a large chocolate labrador. The aim was for him to observe just the basics, such as coming back to me when he was called, and sitting on demand. We'd had two labradors of our own for about two years, and had taken them for training, and while it hadn't made them entirely reliable, it had certainly calmed them down and they would mostly respond to our commands. As Buster was going to be with us for a month or so I felt I'd help out by trying to calm him down a bit too, and getting him to obey the basic commands.

One afternoon I was walking him in our small field, which at the time had no animals in it as we were leaving it fallow. He was actually doing quite well, walking to heel on demand, and sitting, and also coming to me from a reasonable distance. Then suddenly he decided that he'd had enough fun with me, and ran towards the next field – and as often seems to happen, a hole in the fence which had not been there two minutes before suddenly opened, and he scrambled through. I was startled, and only just moving in pursuit as he ran round the orchard fence and towards the goat hut... and the chicken ark.

Knowing that the chickens were out, I started to panic – and sure enough he went straight for them, and they scattered in a cloud of squawks and feathers. Bantams are pretty fast, and all of these could fly a short way, but Buster was fully grown and had both speed and a tight turning circle. I chased after him and tried to dive on him, and nearly managed to get hold of him, but he slipped round me and grabbed one of the yellowy-orange chickens out of the air – and with one quick flick of his head, it was dead. Finally I managed to grab him, and holding him by the scruff of the neck I made him sit, and growled at him. I then told him off severely so that he'd know he'd been a bad boy.

However, it was clear that my telling off did not entirely get through to him, as his tail was still wagging and he was clearly happy with himself. After I'd got him back into the house I went back out and checked the chickens: fortunately he'd only managed to get the one, and the rest were clustered round the ark. I buried the chicken near the place we'd laid Sonny to rest, and then berated myself for my foolishness.

Strangely the incident did seem to have a good impact on my relationship with Buster, because afterwards he was much more responsive to me, and would come to me, sit, and sort of stay to heel (he couldn't quite stick to it, as his excitement would often betray him and he'd run ahead, wagging his tail madly). Clearly he'd decided I was the alpha dog – but I just wish we hadn't had to lose a chicken for him to learn it.

Shortly after this episode we moved the chickens into a half of the orchard that we'd tried to make fox-proof. We put up tall fences, and dug them into the ground about a foot, and put some barbed wire around them, reasonably confident that this should help reduce the fox threat. Although we'd been lucky and not lost anything to a fox (apart from a two-legged one), our neighbours hadn't been so fortunate and had now lost three separate groups of birds, so we were worried that soon it would be our turn. We also had a new gate put in for this area, but unfortunately it had a gap below it.

On the very day I was planning on fixing this gap under the gate I took our own dogs out for training. At this stage I was taking them one at a time, walking them to heel, making them stay and then come to me, and similar exercises. One of them, Jack, was very obedient and responded well to commands; the other, Darcy, still sometimes thought she was top dog, which could make it frustrating as she would take off and disappear from time to time. Still, she was getting better, and that day I was very pleased with her. I was about to take her in, and had made her sit and stay while I walked towards the house; then I called her to come, and she started coming towards me – but all at once suddenly veered away from me. She headed straight for the chicken area, and quick as a flash was under the gate.

Once again the chickens exploded in a flurry of squawks and feathers, with me chasing after the rampaging labrador – but sadly once again the dog managed to grab one of the chickens, Cassie, and with a quick flick of her neck had killed it. I severely castigated her, but just like Buster, she was so pleased with herself I think it had little impact. Of course I fixed the gap under the gate immediately after, and since then have been obsessive about knowing where the chickens are when I am taking the dogs for a walk – but it took me a long time to forgive myself. Once was clearly an accident: twice smacked of carelessness.

Success at Last!

After our first attempt at using eBay eggs you might have expected us to give up, but we were made of stronger stuff. We decided that as long as we were controlling the red mite, and as long as we responded instantly to any other potential problems, it was a good way of getting some additional chickens. With the two we'd lost to the dogs we were down to five, and we still wanted around eight or nine. So when two of the chickens went broody at the same time we leapt at the opportunity, and bought six more eggs in an auction, again a combination of different Pekins. The eggs arrived quickly, but this time we left them for most of the day before putting them under the broodies in the evening.

We knew it would be a month before anything hatched from our eBay eggs, but we checked on the broodies very carefully every day. We made sure they were getting enough food and water, and at least once a week we'd check them for red mite. This was a rather undignified process for the hens, as we had to lift them gently off their eggs and then either turn them upside down, or hold them up high and peer through their feathers. Actually this procedure might not really have been required, but that's what we did! Fortunately our judicious use of red mite killer and powder during the summer seemed to have kept their numbers down, and throughout the whole month we didn't see any live mites at all.

The final few days were of excitable anticipation, as these would be the first creatures actually born in our care. On the twenty-ninth morning after placing the eggs under the hens we came out to feed the animals, and were greeted by the amazing sight of two new balls of yellow fluff sitting in the nest box, with the proud hens mothering them. Over the course of the following day, two more chicks were born, and all four were a joy to behold – when they were visible, that is! They tended to hide underneath their 'mums', who accepted them as if they were their own. On at least a couple of occasions I'd be worried we'd lost one, as I couldn't see it in the nestbox, so I'd then pick up the hens in turn and put them in the main part of the ark to allow a clear view of the nest box – and very often there would still be only three chicks! Then I'd hear a little cheeping and see that the hens had one of the chicks with them – when I'd picked her up I'd managed to catch the chick too...

The little chicks were simply joyous. In the first week or so we'd check on them four or five times a day, just to see them. But they grew up fast, and within a week were climbing around the inside of the ark, and within two weeks were occasionally wandering downstairs and starting to explore their run. We continued to delight in them all through the summer and autumn, and they grew into fine-looking hens and one cockerel – a happy circumstance indeed, because if they'd all been cockerels it's unlikely we'd have had the will to kill them, so we were fortunate not to have to worry about the matter! They were born fairly late in the year, and we knew the hens would be unlikely to lay eggs before the winter, but it was something else to look forward to in the spring. And the broody mums were so good with the chicks, true mother hens!

They started to lay again about three months after the chicks hatched, but obviously slowed down as we headed into winter.

Overall this episode was a great success. We hopefully also had some lambs on the way, but the confidence gained from the chickens started us thinking about the most challenging of the animals we had planned to breed from: the cows!

THE TRAVAILS OF COW KEEPING

Owning cattle is all about bureaucracy, especially with a rare breed like the Dexter. With a combination of the Dexter Society (who generally provide useful information) and the various Defra entities (who provide a quantity of information of variable quality), we seemed to have paperwork coming out of our ears. The first bits of paper are of course the movement forms, which are required for most movements of farm animals, and certainly all the types we would do. Then there was the passport, which contained, per cow, the details of when it was born, and its movements. Goats don't need this passporting system to follow them from owner to owner, which seems rather unfair on the goats – surely their history is relevant and interesting to each new owner? Then there were all the requests from the British Cattle Authority, or whatever they're called, which seem to ask the same questions again and again in slightly different ways, as if to try and catch you out.

These were a small pain for us, but I cannot imagine how a farmer with a hundred head (or more!) of cattle copes with it all. I assume most of them just get used to churning through it, though I fear many will be caught out by some seemingly inconsequential mistake in the paperwork and be swamped by armies of faceless grey men (and women), trying to tease out even bigger flaws! I think we received a letter a week for the first few months on some aspect or another; the sheer cost of postage alone must be staggering... Ranting about Defra is common to most farmers, but those with cattle, and most especially dairy farmers, deserve the right to it most. The above is a pale shadow of a real rant, and fails to describe the true horror of it all. I am sure it is well meant, but much of it really seems to serve no purpose at all...

The Breeding Plan

As mentioned before, the plan with the cows was to get them pregnant, keep any girl calves, and castrate, fatten up and then eat the boy calves. Once again, being a boy on a farm is rarely a good thing, and boy cows have a particularly rough ride – but then, how many bulls does anyone need? In the wild presumably the boys would be whittled away by fighting, or be left out of the herd anyway.

It would appear that there are two ways of getting cows pregnant: either get a bull to 'cover' them (the only apparent use for a male animal), or use artificial insemination (AI). Of the two, AI is the more expensive, but generally considered to be more reliable, and much more controlled. We looked into both options to make sure we were selecting the right course of action.

When using a bull it is usual practice (at least it seemed from what I read) to transport the cows to the bull, and not the other way round. This didn't appeal to us, given our original experiences of transporting Wrath and Avarice – we actually could conceive of no way to get the two of them into the trailer without building some farm buildings and gate control measures, which was unlikely to happen very quickly! As mentioned before, having a bull on our land with the footpath was not possible, so would require separating the fields further so that there was an area without the footpath. To avoid having to spend a lot of money deploying new fences we felt we must look more carefully into AI as a solution.

Not knowing the process for AI – well, obviously we knew it intellectually, but not the details – we called out the vet. I don't normally like calling the vet, but we didn't really have a choice if we were going to progress our plans to breed from the cows. So Jenny, the vet, came out to check the cows with AI in mind. Despite not being able to get closer than about ten feet, she pronounced them to look sound, not too fat (which Dexters apparently have a tendency towards, being greedy little minxes) and with no obvious problems. I was using some food to entice them forwards to allow Jenny to watch them. They were a bit nervous so were sidling around but still trying to reach for the food; this is a strange little dance I often have with the animals, in that they really want the food, but don't want to come too close in case we do something, so they sidle around, and then dart in and out. With the sheep it's quite funny, with the cows the darting is a little worrying, though they tend to back away more quickly than the other animals. As she was being entertained by this spectacle, I asked Jenny how AI might work.

In essence the process would start with the vet giving the cow a cocktail of drugs and hormones to force the timing of her menstrual cycle. A week later a follow-up would be required, and then two days later the man with the sperm would need to turn up and perform his function, and normally he'd be along the next day too, to increase the odds of conception. At least I think that's the process, because once she started to describe it, all I could think of was how many times we'd have to get the cows still enough to have these various no doubt painful injections and intimate (man or woman with sperm) activities performed upon them. As they finally decided they'd had enough of all this and disappeared off into the distance, occasionally kicking up their heels to show their disdain, I realized it might be a significant challenge, perhaps more than that of getting them into a trailer? On balance, however, it seemed as if it would be better to go the AI route and only have to control the two cows on our own land, but it clearly wasn't going to be easy.

The Cattle Crush

Now the challenge was to work out how we would get the two cows into a position where we could properly control them. After spending some time imagining being dragged around the field hanging on to a halter, or watching Wrath either jump over some more pallets or smash a fence to smithereens, we came upon the idea of getting a crush. Instead of the wooden crush we'd seen Wrath demolish we determined we needed a metal one. Once again we went to the internet to see what was available. After railing against websites which refused to publish any prices, or prices of the items I wanted, or prices in a real currency, or describe the product in an understandable way, and so and so forth, I finally found a metal crush on eBay at less than a third the normal, as far as I could tell, price.

My usual experience with eBay is that anything which starts off that cheaply then usually ends up over the odds, so I watched it. Perhaps there was something wrong with it? I checked the description, and those of others on the market, and it sounded fine, it had a 'three-point linkage' which the others didn't always seem to have, and there was implied benefit in this, though unfortunately it wasn't at all clear to me what it actually meant. The days ticked by and there were no bids, until finally it was a few hours to go, and wondering why no one else was going for it, I put in a bid. Then 'bong!' (a traditional eBay sound of victory), it was mine. Much excitement ensued – in my head at least: this was a much bigger deal than a few chicken eggs. I contacted the company, speaking to a lovely Welsh lady who helped sort all the delivery details out, and within a week it had been unceremoniously plonked halfway up our driveway in the small field!

The crush was a large metal box with lockable slam gates at either end, and a lever system on the top which allows someone standing outside the crush to bring internal doors closed around the animal's head to keep it still. It was large, steel, very heavy, and in entirely the wrong place. So Alex and I decided to move it. First we tried dragging it. No chance. Then Alex hit upon the idea of using fence poles as rollers, *à la* the ancient Egyptians (though of course they used special rollers for the Pyramids – which is not to say they couldn't have been used for fencing at a later juncture...). So we got a dozen or so poles which were waiting for us to take the time to hammer them in for some more high quality fencing, and started setting them up in a line in front of the crush. Holding the damn thing high enough to get rollers under it nearly killed me, especially when half of it was on rollers and it was more interested in moving horizontally rather than up (at this point I recalled that the ancient Egyptians had employed a significant amount of unwilling labour, an option not available to me). Still, we'd got about thirty feet along the field, and were quietly congratulating ourselves, while mentally adding up the hours required to complete the move, when my neighbour Rob turned up in his tractor and offered to help. What a gentleman!

In mere minutes the tractor was attached to the 'three-point linkage', which turned out to be three short bars on the side of the crush which matched up to the points on the back of the tractor where ploughs and suchlike are attached. I suddenly appreciated why so many of the adverts seemed excited by this feature, and realized that without it we'd have been forced to use the tractor to drag the crush, which would have seriously damaged the field. It was then lifted slowly into the air – the first lift being a little fast, so much so that it had nearly toppled the tractor – and we were off, and ten minutes later we had the crush where we wanted it, in the middle of the field but much closer to the feeding area. Perfect!

The plan was to get the cows used to the crush slowly, and then when they were fully happy we'd trap them in it a couple of times, feeding them to show them it was OK, until they were used to it, and then set up the AI programme. It was going to take time as the cows weren't stupid, also they had clearly seen something similar before, and Wrath, for one, was clearly unwilling to go anywhere near the crush at first.

Then to make things a little more interesting, just around the time the crush arrived, we received a letter from Animal Health telling us that as we had a new herd, they needed to be tested for bucephallis – (?) though I couldn't see why checking to see that they weren't Alexander the Great's horse was that important – and bovine TB. Apparently it didn't matter that they had been tested the week before we got them, it was required of all new herds (you may recall my earlier observations – this kind of lunacy is rather common). The government pays for this, so I guess I shouldn't complain, but they gave us a short deadline to complete the test, *or else*. I probably should have found out what the 'or else' was, but I assumed it would be something dire. We arranged for the vet to come out in the last week before the deadline, and in the meantime set about getting the cows used to the crush.

When Avarice realized that if she wanted to taste goat mix again she would have to go in it, she was relatively fine with the crush; she would come all the way in following the food, and on occasion even went in of her own accord. Not so Wrath, who was more than sceptical: this contraption was clearly not good and she didn't like it. But several weeks of coaxing eventually got her to the point of entering the crush. The process was not helped by the goats, however, who would rush into the crush to get to the food, thereby preventing the cows from getting to it, or putting them in a position where to get either one of them all the way in would require her to be almost on top of a goat, an unintended crush. . .

After some goat wrestling I managed to get the goats to go around the back of the crush so they could put their heads through the gate at the end to get to the goat mix, thereby ensuring they weren't in the way. When eventually Wrath did start to go in she was fine if she could see me in front of her, but if I walked around behind her, she'd hastily back out. This was a bit of an issue as I needed to close the gate behind her to get her properly secured. With a bit of experimentation I worked out that if I used some baler twine I could set up a pull to slam the door behind her while still standing to her side. With a little confidence I awaited the arrival of the vet.

In the meantime Animal Health had decided that the county was free of brucellosis (not bucephallis, as I'd somehow come to believe) so we didn't need to test for that, which was nice, but unfortunately bovine TB was still required.

The TB Test

The fateful day arrived and the vet turned up, a man called Paul who seemed rather bemused by our set-up. I explained that the cows were a little nervous and asked that he stay by the car while I got Wrath into the crush. Unfortunately she's a smart cow, and the arrival of a man in a Land Rover must have triggered some kind of alarm bells as she was significantly more skittish than usual. Still, I persevered and nearly got her into the crush, but alas she just wouldn't go – and then Avarice pushed past and was all the way in. I was getting rather desperate, so decided (against my original plan) that I'd at least get Avarice, and we could look into Wrath next. As soon as the crush was closed I made sure to get Avarice's head secured so she wouldn't hurt herself and twist around in the crush, and then signalled to Paul to come over.

Wrath went mad, jumping and kicking her back legs a few times before deciding that this was all too dangerous for her and running at full pelt down the field and then down into the dip by the stream so she was out of sight completely. It seemed her fear of vets was significantly more than her fear for her daughter, and her maternal instincts were drowned out. I rather optimistically thought I might be able to persuade her back after the vet was done with Avarice. The testing process was simple, but more than I thought it would be. He shaved a patch of fur away on the side of her neck and then injected her with one of those multiple needle things, the one they use to check for allergies, which I guess made sense.

We then let Avarice out, who was not happy, but not as mad as her mother had been. I then went off to try and coax Wrath back. No chance. She wouldn't even let me get within twenty feet of her, she was that upset.

Wearily I trudged back and apologized, and wondered if he could come back to do the test in a week or so on Wrath. He then hit me with two unhelpful surprises: the first was that the government usually wouldn't pay for a second visit, so I might have to, and the second was that really all cows on a holding need to be tested at the same time. If I'd known that I'd have tried for longer to get Wrath in the crush . . . but that was water under the bridge. I asked him what would happen if we didn't get them tested, fearing the worst. The sanction would be that we would not be allowed to bring any more cattle on to the holding, or take our two off, until we got them tested. Another surprise, as I felt that with a bit more time I would have got Wrath more acclimatized to the crush and been able to catch her. . . It shows that it pays to ask more questions and not make assumptions, especially when officious documents arrive in the post.

Paul left, still bemused, and promised to return in three days to check the results on Avarice, because at least that would tell us if we did have a problem, even if it wouldn't tick the box with the authorities. I definitely didn't know about the return check-up trip, as I would have had no confidence in getting the cows into a crush twice in three days.

He also told me there was a farmer a few miles down the road who either couldn't or wouldn't check his cows for bovine TB, but because he had no intention of ever moving them off his holding, or indeed of getting any more, it didn't worry him. And, more importantly as far as I was concerned, Defra and the other cattle authorities seemed to be accepting that position. I assumed that if there was an outbreak of bovine TB in the area things would be different, but it did put my mind at ease – after all, the letter had said we had to complete the test by a date which was fast approaching.

A few days later he returned. Despite deploying my devilish charm and a significant amount of goat mix, I had not been able to coax either of the cows within thirty feet of the crush, so asked if he'd be able to check the injection site without Avarice in a crush. He said that as it was only to see the result of the test, that should be fine, so he walked with me to where I fed them, and for the last few metres I went ahead of him and then poured the feed out. The cows cautiously started eating, but as soon as Paul stepped nearer they were off. Fortunately he'd managed to see the side of Avarice's neck, if only briefly, and he assured me she did not have bovine TB – which didn't surprise me, given they'd been tested so recently!

All of this excitement established clearly in my mind that with our prevailing set-up we had precisely zero chance of performing successful AI on the cows. Even worse, we now couldn't go for the bull option, even if we fenced off an area, as we'd still need them tested for bovine TB first. This was all very depressing, and it remained the state of affairs for some time.

Noisy Cow Nuisance

Another source of concern for us was winter feeding for the cows. Our first winter with animals we'd only had Bill and Ted, and while we fed them twice a day instead of once, and we spent extra time on the water, little extra effort was required. I think we may have bought a couple of bales of hay, but it was all rather ad hoc and the goats seemed fine. The second year we had five goats and two cows (even if Avarice was still only young), and food became a little more of a burden. We'd provide them with hay twice a day, supplemented with goat mix in larger quantities (two or three saucepanfuls). However, it wasn't all plain sailing.

As the nights closed in and we started having regular frosts the grass was providing no real food value, which meant that Wrath and Avarice would wait for us at their feeding times, around six in the morning and seven at night. And while they waited they mooed, loudly, furthermore maintaining their mooing until we had delivered their food, however long that took. Not only did this impact any thoughts of a lie-in on the weekends, but they were so loud they could be heard from the village pub, some half a mile away. We were worried that while people might like the sounds of the country, they might not want quite so much, or quite so early in the morning. In particular we worried about our neighbours, as the cows were effectively bellowing at their windows. Their youngest even started to have fears of the cows climbing up the stairs to get him – though he loved them in the daylight and was happy to visit them!

Our solution was to feed them a bit more, but we were now stuck as they seemed to be getting a little fatter, and yet still didn't stop mooing. Some experimentation with volume got us to the stage where they would only moo a bit (well, still quite a lot) while they seemed to stay roughly the same weight. This is judged by the shape of the top of their back, where it should, I believe, be possible to make out their backbone and the curves of their rump on each side, without it all being subsumed into one single rounded curve. For the following winter we resolved to move them to a different field, where at the very least they wouldn't be mooing directly outside people's bedrooms – but for that we needed to do quite a bit of fencing first.

Fine Wines and Fencing Vines

One of my hobbies is drinking wine. Now this may not seem a hobby, but I've been on several trips abroad solely to taste and try out wines, and I invest a lot of time and money into wine, to the extent that it is more than just a casual pastime, therefore I think I can safely describe it as 'hobbyish'. Alex just likes to drink nice wine. For several years we'd talked about retiring (we like to plan ahead), and one of our retirement plans was to own a vineyard, possibly in northern Spain, as we both love Rioja. As part of the plan I decided I really wanted to grow some grapes to get some experience of them.

First we needed an appropriate site. Our fields all slope, but sadly in a northerly direction, which made them less than perfect. However, we had a lot of rubble from our conversion, and our neighbours had even more from their renovation, and so an idea was born: we would build our own south-facing slope at one end of our field using the rubble. Builders in big dumper trucks very quickly built a nice hillock for us, and even made the south-facing side into a gentle slope with the top covered in soil. However, because this was all being done during the winter, unfortunately they also made a real mess of the small field, as they created a muddy path to the new slope. Nevertheless I felt this was a price worth paying to get the slope.

Alex and her mum very kindly purchased some young grape vines for me for Christmas (in the sense of ordering them for delivery in the spring). I'd asked for two white varieties and two red, to see how they would grow in our environment, and waited with bated breath for them to arrive; there were twenty-five plants in all, and when finally they came I rather excitedly planted them in five rows, the recommended two or so feet apart. I was inordinately proud of my vines, especially when in early May they all showed signs of growth and some even had fully fledged leaves.

When we'd put the slope in we'd also noticed that the fence on that side, which was intricately tied into a hedge and bordered the road, was actually relatively feeble – only three strands of sagging barbed wire. We knew we needed to do something about that to avoid having intrepid goats in the road, and so some fencing was in order. Given that we also didn't want any animals mixing with the vines, based on their tendency to eat anything green and deemed valuable by me, we needed to protect those, too. The easy option was to fence off that whole section of field, neatly solving both issues in one hit. The fence was tall mesh, which we knew would keep the goats out, and we were confident we'd have few problems. However, we had failed to take the cows into account.

All our animals, and I can only hope that this is a feature common to farm animals in general and not a personal curse, like to test fencing. In Wrath and Avarice's case we'd only seen evidence of them going through pre-made holes, so they had given us less cause for concern than the real escape artists – the goats, and to a lesser extent the Soay ewes. It's true that in this case the cows didn't actually make holes deliberately, but what they started doing was leaning against the fence, particularly near the south-facing slope (perhaps it gave them shelter from the wind), and slowly this caused areas of the fence where we'd joined up two sets of mesh to stretch, and then be dragged open. Whereupon the cows would suddenly notice the new hole, and innocently wander through to investigate, closely followed by the goats, there to feast upon a wealth of greenery – unfortunately including my precious vines!

This happened three or four times over the course of a month, until I double-meshed a whole area of the fence, which seemed to put them off for a while. Some of the vines never recovered, and the others were feeble and timid affairs, perhaps wondering why they should put in any effort if they were going to end up as fodder. I have hopes they will recover in a year or so, and even be stronger as a result of the experience. Next time I have something new I particularly want to protect I shall be far more careful, and probably put in two layers of fence to provide defence in depth!

Always the Optimists!

The travails of the crush and the fencing left us in a bit of a pickle with the cows, but we knew what we needed to do to resolve it, and that it would take time: some extra fencing and some efficient control measures, and all would be nice and easy (when the facts look dire, blind optimism will see you through!). Having cheered ourselves up with these thoughts, our minds turned to another challenge we'd also recently decided to accept: a new set of pigs!

PART 3: EXPERIENCED HANDS

EXPANDING THE PIG HERD

When our neighbours gave up sheep, they also gave up pigs. We were fresh from our successful experience with the OSBs so were not unnerved by taking on some pigs again, but this time we were going to be 'everythingers': we were going to manage the whole process! As part of the deal – which involved a small sum of money and the promise of a finished pig in the future – we acquired a boar, Humphrey, whom we knighted Sir Humphrey after the character from *Yes, Minister*; a sow, Polly, whom we renamed Persephone – or at least I did, allegedly because I think that long words make me look smart – or Percy for short, in part because we love the sweets from M & S; and six of Polly's piglets, three boys and three girls, who were about three weeks old. All the pigs were British Saddlebacks, an easily recognized rare breed (though amongst the least rare) being black with a thick white stripe (the saddle) around their back and front legs. They are lop-eared, which means their ears fall forward over their eyes, and they are supposed to be easy-going. At the very least we were hoping they would be rather gentler than the OSBs had been.

Bringing the New Pigs Home

The enterprise didn't start off in the best of ways as we had our first taste of what it means to 'squeal like a pig'. If you ever have the chance to pick up a piglet, it's probably best if you don't. We found that if we were to pick up or attempt to pick up one of these piglets, it would scream like a banshee, though it would stop as soon as we put it down again. The sound was so loud and high pitched it was actually painful. In our early days of pig keeping I'd heard that the health and safety people had mandated ear phones for farm workers dealing with pigs, at which I had scoffed. Our OSBs had been fairly quiet, even when we picked them up as piglets, and we'd never heard much from next-door's pigs – but in a few short minutes with these Saddlebacks I was educated, and withdrew my scoffing immediately!

So the scene went something like this: 'IT professional' and City lawyer, already slightly covered in pig poo (which seems inexorably attracted to clothes – it almost flies up and sticks of its own accord), start trying to entice seven piggies into a trailer – the sow and her six piglets. So how hard can this be? Based on the sheep we had thought 'Let's just throw the food in the back and they'll walk in'. This didn't seem to work, however, probably because the piglets were still quite small and not much interested in solid food, and the sow wasn't happy about all the people around her. This was followed by a little pushing and more direct encouragement: the slap on the bottom, variations of goose-wing flapping, and the occasional 'pick up a piglet and be deafened' technique.

We slid around the pig pen with our neighbours, alternating between cajoling and pleading. As the attempts to look like we knew what we were doing rapidly vanished, the pig poo and tempers started to rise. It was really like trying to herd screaming banshee kittens. 'Are you sure these aren't your progeny?' was my final comment, to which Alex responded, 'Am I supposed to be the ... pig in this scenario?' Eventually we had all but one of the piglets safely in the trailer.

This last piglet looked rather like Quasimodo, with a massive – and I mean massive, a third of its own size – hump swelling on its back leg. The week before when we'd first agreed to take on the pigs, we'd seen the piglet, and Rob had said he wasn't sure what the issue was, but that we didn't need to take that one if we didn't want to. I decided to research the issue, and discovered a potential solution: according to my sources the likely reason for the bump was that the piglet had suffered a scrape and its body was fighting it, and one of the ways it did that was by forming such a lump – it was likely that it would be filled with liquid and pus. All we had to do was lance it, and all should be well.

So armed with only a big needle and a love of pus we set to work. Holding the piglet carefully, its squealing subsided a bit, and Claire pushed the needle gingerly into the hump, and out came ... clear fluid – lots and lots of it, it probably took five minutes before it all came out. No blood, which was a good sign as it meant it was probably not infected, and no pus, which was also good because that much pus would have been pretty disgusting. After which the lump was almost completely gone, and we quickly put the piglet in the trailer with the others. In fact the treatment worked so well that a week later it would have been difficult to identify which piglet had had the hump, so completely did it heal. We breathed a sigh of relief that another vet's bill had been avoided, at least for the time being, though Rob's threatened alternative had been to use his shotgun ... so at least we saved the little snuffler from that unhappy end!

Moving Humphrey was by comparison a dream: we opened up the trailer, waved a bucket of food at him and he ran towards us. We poured the food at the back of the trailer and as he rushed in to get it we moved smartly out of the way and closed up the trailer ramp behind him. If only it could be that easy every time. But more on Humphrey a bit later...

As with many of our endeavours, we hadn't entirely planned it all out. Familiar anticipation at caring for some new animals was therefore also accompanied by the dread in the pit of our stomachs that we didn't really know what we were doing, and that things might go horribly wrong.

Fortunately we still had the pig hut in the orchard where the Oxford Sandy and Blacks had lived, so we could keep Percy and her piglets there, but we needed another space for Humphrey, and a bed for him for the night-time. As the piglets were so small it didn't seem sensible to have them sharing with a large boar who might crush them or otherwise hurt them. Our neighbours kindly lent us their hut, which of course they no longer needed, but didn't want to sell in case they decided to try their hands at pigs again. We hastily fenced in an area in the centre of the field; as I recall it was chucking it down with rain, as it always seems to when I want to do some fencing, and I ducked back into the Land Rover every ten minutes or so to try and warm up and to eat a couple of pieces of chocolate. Once the area was done we were as ready as we felt necessary, and that's when we moved the pigs over.

We should pause to consider the fencing at this point. I've mentioned it a few times before, but fencing is such a critical part of looking after animals that when I think back to how naive we were, I cringe. Our fence posts were pounded into the ground by hand, which meant that if we were tired, or had a lot to do, at least a few wouldn't be completely solid, and most would be at an angle. We still weren't getting the mesh tense, though to be fair we were getting better, in that I at least tried to pull the fencing tight while Alex nailed it in. It's almost as if we trusted our animals not to test the fence, as if we knew they wouldn't want to make us look bad, or go somewhere without our permission. Actually I think they were probably smirking at the thought of all the fun that lay ahead!

Percy Falls Ill

Percy and her piglets settled into their new home quite comfortably and all seemed well. The water was set up properly, we had a little green tub which measured out the correct amount of food, and they had plenty of greenery to snuffle around in. We fed them each morning and evening and made sure their water was working all right, and everything seemed well; but a few weeks after the pigs had moved in, my father-in-law mentioned that he thought Percy was looking a bit thin, and looking at her with fresh eyes we realized this was true, she was starting to look rather hollow.

Obviously I was worried we hadn't been feeding her enough, but she also seemed to be becoming less interested in food. My theory at the time was that the piglets were sucking all the energy out of her, although this didn't explain the loss of appetite, because surely then she'd want to eat more, not less? The piglets were about ten or eleven weeks old at the time, so were ready to be weaned. The problem was that we didn't really have anywhere for them to go other than with Humphrey, which wasn't yet something we were willing to consider.

After some discussion, we agreed to get the vet out to visit (despite my concern about the cost) and see if she could help. Her advice was that the piglets should be separated as quickly as possible, and that we should make sure Percy ate as much as possible. She also injected her with antibiotics and vitamins to give her a boost. But her basic point was that her depressed state wasn't too much to worry about, as sows tend to get worn down by all that suckling, and when she had the chance to focus on food without six increasingly large piglets to compete with *and* then feed herself, she should be as right as rain before too long.

We were rather confused as to what to do, as everything we'd read and heard emphasized that pigs are social animals, and we didn't want Percy to be lonely. Then again – and this is the second law of the smallholding – never believe what you read in the books. (The first law being plan out and then pay someone else to do your fencing, no matter what it costs – and it does cost a lot: nevertheless get someone who knows what they are doing, and get them to do it all.) Books contradict each other, sometimes within the same book, and never replace following your intuition or, better still, asking for advice from someone who knows what they are doing and following *their* intuition. After some debate we decided to move Percy and leave the piglets where they were, because we could then build Percy a shelter out of straw bales, and she could be in the field next to the piglets so she could see them.

Percy had a bit more energy, but continued to ignore her food. We bought lovely Braeburn apples for her, which she ate for a while, but then she lost interest in those, too. She was listless and, to us, appeared to be clearly unhappy. She was still drinking though, so I got some apple juice and put it in with her water and she seemed to love that. Alex then went and got some more, making sure it was the best organic apple juice. Definitely no concentrates for our sick piggy! Percy summoned up enough energy to move in with the goats, but then seemed unable to move any further, and the goats, suddenly homeless as there wasn't enough space for both them and Percy, were rather unhappy. It was fairly cold March weather, not freezing, but not pleasant, so we were glad that at least she was under cover.

We were now getting very worried about her. It's difficult to explain the feeling of frustration and helplessness we were experiencing. Vets often say that treating animals is much harder than humans, because at least we can say where it hurts! We decided to get the vet out again (I didn't even protest at the cost this time). It was a different vet this time, and her advice was that Percy was in a 'negative energy spiral', whatever that means, and what we needed to do was 'gee her up' and that would get her eating again. She also recommended that we move her regularly, in effect to flip her over from one side to the other. The words 'bed sores' didn't actually leave her lips, but that was the implication. Another shot of vitamins, and the vet was off, leaving us somewhat mystified (and angry) in her wake. It appeared that our really quite large sow, which hadn't eaten properly in days, now needed to be given nothing more than a pep talk each morning, flipped from side to side throughout the day, which was no mean feat despite her having lost a lot of weight, and a multivitamin. The vet had been quite charming, but at £200 and counting we were hoping for a little bit more than 'gee her up'. Not really knowing what to do, we continued with the organic apple juice, and hoped.

The next day Percy was dead.

This was very upsetting. It was our first big animal to die unnaturally, an animal we'd only been looking after for such a short time, and (heartless as it sounds) two expensive vet visits into the bargain. Alex phoned the vet to ask them how to dispose of the carcase, and they responded, not with sympathy and advice, but with an attitude that as the pig was dead it was no longer their problem! Alex therefore spoke to Animal Health, who were very helpful and provided the details of what we needed to do. On their advice Alex then phoned a man to come and pick up the body (he had a large truck with dead animals in it, such as horses and cows) – a job that would harden anyone's heart – and importantly, to leave us a document saying he'd done so. After this incident we decided to move vets, after a letter of complaint and a half-hearted apology. We also decided we'd call the vets out less often, as our faith in them had been further shaken by this incident.

As a coda to the story, Alex was rather annoyed and upset with me a week or so later when we were talking about Percy. We were still worried that we'd missed something, or that there was more we could have done, and I mentioned that Persephone was a Greek goddess who was married to death and lived in Hades for six months of the year (thereby causing autumn and winter). Apparently I should never have suggested such an ill-starred name...

In hindsight, and with what we've learnt since then, I think we could have made Percy's last days more comfortable, but I don't think we could have made her live any longer. It's astonishing how little we – as in 'mankind' – know about animals, even those we've domesticated for thousands of years. There is a vet's saying: Sick Sheep Seldom Survive, meaning in effect that once a farm animal is sick enough to show symptoms, it's probably going to die. Knowing that now surprisingly makes it a bit easier to cope with the deaths of animals, though it is always sad.

Saying that, they are phenomenally resilient as well, as Bernard's story will show...

Two New Sows

A month or so after the death of Percy I resolved to get two new sows. The plan was to alternate them with Humphrey so he always had a companion (lucky boar!), in the hope that this would reduce his wandering ways. He was ignoring our fencing at will, and we'd often find him visiting the goats or frolicking with the cows. I found an advertisement on the internet for two British Saddleback sows for sale in Buckinghamshire, so off I went with my father-in-law and a trailer to pick them up.

The farmer who sold them to us had been farming for forty years and in the past had looked after many hundreds of pigs. He wanted to make sure we had experience of pigs before he'd sell to us, and I assured him we did. Apparently a year or so before he'd sold a couple of pigs to a city-born fellow who'd moved out to the country. He'd sold them as weaners, so they would have been about nine to ten weeks old and twelve to fifteen inches long. A few months later he had a call from the unfortunate man begging him to take the pigs back as they were destroying his garden! He hadn't really considered what would happen when the weaners grew into real pigs.

So after explaining we'd had pigs before, both OSBs and now British Saddlebacks, and a chat about what a nightmare Defra were, he showed us the pigs. One was a perfectly normal Saddleback sow, which looked in good condition. The other was a bit more unusual, being almost totally white, with a black snout and a black rump – though the farmer assured me she'd give birth to normally patterned piglets. He also revealed that it was his daughter's favourite pig, and that she'd begged him not to slaughter it; but he'd decided the time had come for the pig to go, so selling her to me seemed the best compromise. (I really wouldn't want to be in his shoes when she found out, but I'm sure he knew what he was doing.) So I gave him some money, we filled out the movement licence, and off we went.

We named the mostly white one Bernard, after the private secretary in *Yes, Minister*, and the other one Hacker, after the Minister himself. We felt they'd both go well with Sir Humphrey!

Hacker was lodged initially with the other weaners (Percy's progeny) and Bernard went straight in with Humph – though this may have been a mistake, as Bernard was still quite young and looked so small next to Humphrey; but it was the plan, and after a few days they seemed to be getting on all right. Then one day we noticed that Bernard was limping and had a nasty gash on her back right leg, the cause of which, we came to think, was some loose barbed wire in the area with the two pigs – it was likely she had caught herself on it while avoiding Humph.

Barbed wire has its uses, but I've never been totally comfortable with it. In Humph's area it was part of some old fencing which I'd left in while adding the new fences, thinking that it might help to reduce Humph's attacks on the fence on that side; otherwise it had never seemed to bother him. But seeing the damage to Bernard, we decided we needed to get it all out immediately. It was tough work with Humphrey 'helping' us, usually by knocking us, or trying to bite me (though never Alex, for some reason). So we got into the habit of taking some pig food with us, running to one corner with Humphrey chasing, throwing the food to the ground, which he'd immediately snuffle for, and then running back to get the wire out. We'd keep an eye out for him, and when he started to head back towards us we'd run to the other corner, with Humph happily galloping along after us, and pour some more food into that corner, which he'd then dive into. This went on for an hour or so the first time, and I think he loved it! It was certainly good exercise for us, but not the most efficient of processes.

Bernard in Trouble

Bernard's knee got worse and worse and eventually she stopped moving, and seemed less interested in her food. This was terrible for us, it was like Percy all over again, with the added twist that her injury was clearly our fault; so once again we called out the vet. This was the last time we used our old vets, but in fact the woman who came out was one of our favourites. We'd sectioned off Bernard in the hut so that Humphrey couldn't get to her, but he was continually trying to break in and had broken through the barricades again when the vet came. She checked Bernard out, and then leapt out of the pen! She was clearly quite frightened of Humphrey, and that perspective allowed us to see afresh his large, tusk-like teeth and the fact that he weighed at least twenty stone... so we started showing him a bit more respect after that.

Her advice with Bernard was to continue to give her water (we'd been carrying it down) and food (she was at least still eating a bit), and to inject her daily with painkillers and antibiotics. Now I'm not great with needles – I don't mind so much if they're injecting something in, but I hate blood tests, and I am not keen on the sight of big needles.

Bernard, however, was not at all interested in being injected. I'd crouch down to get into the pig hut where she was lying, and initially she'd be fine, but as soon as I put the needle in her she'd be up and moving. It was a fairly scary experience, given the size of the hut and how big she actually was – she probably weighed thirteen stone or so at this point. In her desperation to get away from the needle she'd even run on her bad knee. I broke a number of needles and was starting to get worried that not only would we never be able to get the antibiotics into her, but that it was actually making her knee worse.

After a few days of this I recalled some advice from a friend at work who'd looked after pigs, inasmuch as they were far more controllable if they had a bucket on their head (and I could see some logic to that). So Alex and I developed a routine where I'd chase her out of her hut, Alex would get the bucket on her head and we'd then back her into a corner and I'd inject her, with both the antibiotics and the painkillers. Amazingly, once she was in the corner with the bucket on her head she would stay completely still, even when we put the needle into her. The needles were about an inch long and they needed to go all the way in, so she clearly felt them, but the bucket somehow calmed her so much that she wouldn't struggle, though she was clearly unhappy afterwards! This method allowed us to complete the course of both sets of drugs without losing another needle.

Within only three or four days after we were at last able to inject her, she showed improvement. She was able to get up more easily, move to her food without too much encouragement, and was getting harder to catch! The vet had said to us that even if the gash healed properly she would always limp, and the implication was that she would now be useless as a breeding sow so we might need to think about getting rid of her anyway – but we didn't consider that for one second, and were just happy to see her improving.

Bernard actually recovered so quickly that a month after she'd been so incapacitated that she couldn't even get to water, she was up and about and competing (if not beating) with Humphrey for food, and we'd been able to stop rebuilding the fencing keeping her separate from him, as she seemed fine with him. Our feeding strategy was that we'd always lay the food out in three or four piles to give Bernard a chance to get her own food before Humph pushed her off – she'd get about thirty seconds at each pile before he came to see what she was eating, and then she'd run to the next one. A few years on and you wouldn't know she'd ever had the injury – she walks perfectly, and is as lively as a sow should be; in fact when food is being offered she'll happily run the thirty metres' length of her pen!

Humphrey

While all the drama with Percy and Bernard was going on, Sir Humphrey was teaching us a few lessons of his own. Lesson number one: no fence can keep Humphrey in if he wants out. Lesson number two: try to avoid letting Humphrey sense something on the other side of the fence that he might be interested in!

Humph is a large and aggressive creature. He's about five feet long, weighs at least twenty stone, has large tusks and is immensely strong. He's also male, which means he is hungry, horny and lazy (no criticism intended – I think he's got a great life!). We'll start with the last of the three: Humph likes nothing more than digging himself a mud bath and lying around in it for hours at a stretch, only stirring himself to turn over. If he's already eaten his fill he'll ignore people entering his pen unless they bother him. He also loves having his tummy rubbed, and if it's done properly he'll stretch out and then just collapse on one side (best to be light on your feet around him!) and lie there in contentment as you rub him, fairly robustly so that he can feel it.

Let's take hunger next. Perhaps 'hunger' is the wrong word, but if Humph sees anyone else getting food, he wants some, and not just the other pigs. Oh no, if he sees the sheep, goats or cows getting fed, then he feels entitled to some of his own. Soon after he came to stay with us he amply demonstrated that my fencing skills were not up to keeping him in, and he managed to make holes in all our internal fences so he could happily move around all our fields with impunity. This meant that when we fed any of the other animals he'd be up there, pushing them out of the way and generally making a nuisance of himself. None of them would get in his way, so he'd happily eat everything they had if we let him. The best strategy was to take his food out first, get him busy with that and then feed the other animals out of his sight.

Pig feed has a relatively high concentration of copper so is poisonous to sheep if they eat too much of it (the copper builds up in their systems and then one day they just die), so it wasn't even possible to distract him in the main field – he had to be in his own area for us to give him pig feed. Therefore we would use ewe feed to get his attention while we enticed him back into his own area. We had to do this often and we spent a lot of time patching and strengthening his fence with boards and stronger mesh. Sometimes this would hold him for a few hours, sometimes even a day or so, but never more...

In retrospect part of the problem was that there wasn't enough to entertain him – Bernard was sick and not interested, and there weren't any other pigs in the area with him. There was also a lot going on around him, with all the sheep and cows and the regular daily feeding routine, so he no doubt felt he needed to get involved!

That brings us to the last of the descriptors – horny. Humphrey, like all young males, has a very healthy libido and would like to be active all the time, food and lazing allowing, that is. In his wanderings around the fields he had noted Hacker and the little pigs – the three boys and three girls all sired by him around seven months previously. The main fence to the small field had been built at the same time as our main driveway fence and was therefore fairly sturdy, having been done by professional fencers with proper equipment, and indeed seemed to hold him off. Unfortunately there was a section of fencing at the corner of the orchard which was rather less professionally implemented, consisting of some mesh and hope as far as I could tell – after all, it was really only to prevent the sheep getting in with the pigs, and we still hadn't learned too much about fencing at this point!

It didn't take Humph long to find this weak spot, and then not long at all to push through it, so that one morning coming out to see the animals I was shocked to see that Hacker had grown massively overnight ... except it was actually Humph looking rather pleased with himself. I, of course, resolved to move him back to his own area, which necessitated the use of verbal encouragement and food – well, mainly food. He happily wandered back to his area, and then I fed everything else. Alex and I then spent half an hour fixing the hole in the fence and nailing a board across.

Later that day we were out in the fields and saw Humphrey trying to get over the board; at first it seemed that our plan had worked, as he couldn't get over it, but then he literally climbed up it and pulled himself over, it was astonishing! Once again we led him back to his area, and distracted him with food, and added a little more height to the fencing. It seemed enough to hold him. The next morning… Hacker once more seemed to have grown, and once again I realized it was Humphrey. This went on for about a week, during which time he paid attention both to Hacker and the three girls. Then he just seemed to lose interest.

Obviously they had all been in season and that was what had attracted him to the area, but once the hormones were no longer filling the air he was happy to return to his own hut. Of course this didn't stop him wandering whenever he wanted, but we were slowly beefing up the fencing, and eventually got to a point where he didn't break out any more. This coincided with us moving the three boy pigs in with him and Bernard, and probably meant he had enough company to keep him happy.

Coping with Expansion!

So now we had Humphrey and three young boy pigs; Bernard, now mostly healed and likely pregnant; Hacker, likely pregnant; and three young girl pigs, all potentially pregnant. With all these potentially pregnant pigs it was possible there could up to sixty piglets on their way! At that point we hadn't had any at all, but our planned slow start to the process had been completely ruined by the amorous Humph. We also had only two pig huts and two pig areas, and this clearly wasn't the best situation to bring all these piglets into. We resolved to get some extra huts and to fence off some additional areas so that we could split up the pigs. One of our biggest concerns was that Humphrey might accidentally squish the piglets while they were still young. All the books seemed to agree that the sows had the potential to squish their piglets, accidentally of course, so we reasoned it could only be worse with a boar.

We ordered the pig huts from the same place we'd ordered our first hut, a website dedicated to selling refurbished huts. The owner is a pig farmer who as a side business buys knackered pig huts from other farmers, fixes them up and then sells them on the internet. It seems to keep him busy and probably provides a steadier income than the pigs, given the very cyclic nature of the pig industry. The huts arrived, looking like old-fashioned air-raid shelters, and we arranged them roughly where we intended to add our new fences. Our plan was to have four areas for the pigs in the centre of our fields, far enough away from our neighbours on that side to avoid any potential complaints – the law stipulates something like fifty metres distance, and our original pig area in the centre of the field was well over that from anyone. The orchard, however, was actually less than that to Rob and Claire, but as they had had their pigs a similar distance from us, we considered it probably wasn't an issue.

We would then rotate the pigs through these areas and the orchard, so as always to have one area empty to allow it to recover and avoid becoming 'pig sick': a condition where the soil becomes unhealthy due to the high concentration of pig poo and urine. This was actually unlikely to be much of a problem for us as we were giving the pigs fairly reasonably sized areas to live in, but it's still good practice to rotate the land!

Now we had the extra huts, all we needed to do was fence them in and we'd be able to split out the pigs a bit more and have much more control of things. But unfortunately it turned out that we were rather busy for a few weeks, and we didn't get round to completing the fencing in time to give the pigs the separation they needed. Pigs are pregnant for three months, three weeks and three days, which is actually a really short time, and before we knew it the first births were upon us.

Actually, we were not even aware of it for a couple of days. Bernard had been looking very heavy around the udders for a couple of days, and all the books agreed that this was something which happened in the days before birth, and that as soon as we saw this we should be ready for the little pigs to arrive. Other than this Bernard hadn't changed much – she still ran to get food and barged the boys out of the way if she needed to. We prepared a new area by hurriedly fencing around one of the new huts and putting in a temporary gate, laying straw in the hut and generally making it nice for her. We then went to move her.

The Sows Start Farrowing

In the course of the move I saw something which amazed me: there was a piglet nosing round outside the hut. Bernard's litter had already arrived! However, our joy was short-lived, because of a litter of six, four of the piglets were already dead, either on arrival, or perhaps having been squished. It wasn't really possible to determine which, and the remaining two were very weak.

We immediately moved Bernard and the two remaining piglets to the new area, but one was so weak it could barely move, so we decided to take it inside to warm it up – the autumn was starting to get cold – and feed it milk, per the emergency instructions in the books. Warming the piglet up involved putting the oven on and putting the piglet in a little tray in the grill compartment which was above the main oven, as this was warm but not too hot. Sadly it, too, died overnight, despite all the energy we put into trying to save it. The last one only survived another day.

We were devastated. If we'd moved Bernard a few days earlier, might we have saved them all? We'll never know. But it taught us a harsh lesson, and we were adamant that we wouldn't make the same mistake again.

We finished off the area in which Bernard was now living alone, bringing a water drinker into it and redoing some of the fence more permanently, but moved her in with Humphrey and the boys. In the now empty area we put Hacker and one of the other girl pigs, while the other two probably pregnant pigs remained in the original area in the orchard. We figured this would mean they all had company, but with less risk of any little ones being squished, particularly by Humphrey.

It was only a couple of weeks before the next litter arrived: it was Hacker's, and of five piglets born, only two survived. It wasn't clear what had happened to the other three, but as I found them just a few hours after they were born, I think it might have been that they just weren't strong enough, or they were stillborn. Still, two were better than none, and they were incredibly cute. Much as with the lambs, I chose not to count them until they had survived twenty-four hours – but survive they did, and they seemed to prosper. It helped that Hacker probably had enough milk for twelve or so piglets, and they had it just to themselves!

A couple of weeks later two more litters were born, with ten and eight piglets surviving from each. The following day the last litter was born, and a further eight piglets survived from that.

Taking the Rough with the Smooth

This was a very emotional time for us, and I realized that livestock farmers are absolutely centred in the circle of life. Each litter had brought a flush of new life on to our farm, but also a little death. We met each new litter with joy, but also with trepidation and a sense of dread, wondering how many piglets we would have to bury. We had twenty-eight surviving piglets, but had lost about fifteen. I buried the piglets in our other orchard to provide food for the trees, give them a decent end, and hopefully avoid them being picked over by foxes or other carnivores.

I was certainly more upset during this whole birth process than I had thought I would be, and even considered giving it all up. It just didn't seem worth the emotional cost of burying the tiny piglet bodies. We'd learnt (or perhaps learnt again) the lesson which is important to all livestock farmers: that if we couldn't take the rough with the smooth then we needed to stop. However, after a few days no more piglets had died and I soon started to delight in those that had survived. They were full of energy, and would frolic around merrily as we fed their mothers, and it was impossible not to smile at them. I realized that it was all a balance, and while we should not ignore the losses, we should also not let them blight the victories, and twenty-eight piglets could be considered nothing less than a great victory!

Rampaging Piglets

Having all these extra pigs was, at least initially, no real extra work. We had to feed their mothers a bit more, but other than that they were no trouble. They tended to stick to their mum's side, and when they weren't with her they wouldn't venture too far from their home huts. It was at this point that we decided to name Humphrey's daughters, though only two of the three. The first we named Malcolm (after Malcolm Tucker from *The Thick of It*, keeping us aligned with the political satire-based names) and the other Snowball (no strong reason I can think of, except perhaps an oblique reference to the Simpsons' cat).

After three or so weeks the piglets were quite a bit bigger (having grown from something which would fit on my hand, to proper little pigs around two to three hands long), and they were a lot more adventurous. One morning when we went out to feed them, a string of piglets ran across the small field to meet us, running through the sheep: they'd found, or made, a hole in the fence, and were now excitedly chasing around the field, disturbing the sheep, who clearly weren't quite sure what to make of these creatures! Piglets start to eat solids fairly early, sometimes after only a few weeks, even though they still suckle for up to ten or twelve weeks before natural weaning. This meant that when I put the food down for the sheep and cows, the piglets were right into it, and this did cause some consternation. I tried to get them to follow me with the pig feed back to their mother, and they did for a while, but never all the way back. Once food is down pigs don't want to go too far from it, probably on the basis that it's better to stick to the food you can get your snout into than follow a man with a blue bucket...

After a couple of attempts on my own I retreated back to the house to gather reinforcements – Alex. I figured that between us we could outfox a group of rampaging piglets. We patched the hole in the fence they'd come through, and opened up the gate, giving the two sows in there, Hacker and the as yet unnamed Humphrey daughters, an extra bucket of food to keep them occupied. Humphrey, who was watching through the fence, was clearly unhappy that they were getting this extra ration, so we had to give *him* a couple of handfuls too, otherwise I think he'd have come straight through *his* fence!

Then we gathered ourselves for the piglet herding. I shall not dwell too much on this, but picture in your mind two allegedly highly educated city folks (as we'll probably always be, even if we spend the rest of our lives in the country!) running around a field at 06:30 or so on a weekday morning, with their arms out wide to help with the herding motion. The grass was wet and the pig area muddy, which meant that our legs were soon soaked, and when we *did* get a piglet into the pig area we ended up covered in mud. In addition, Alex discovered the awesome sucking effects of thick mud as her wellies got stuck and she ended up flat on her back in the middle of the pig area with her feet waving in the air.

Showing sympathy, I doubled up laughing, and then intimated that wallowing in poo while there were pigs to be caught was not the optimal strategy – upon which she forcefully told me where to go. She finally managed to get up again and was soon walking barefoot through the muddy field trying not to think about the poo and the other things squelching under her feet! After twenty minutes or so we had managed to get all the piglets back into their area, except for the last two, which we had to corner and carry back with them shrieking and squealing all the while. We then had to go to work, though I for one felt like I'd already put in a full day.

Fencing: a Continuing Challenge

Over the course of the next few days this was repeated several times, until the weekend arrived, when we could do something about the fencing. We'd thrown up the fencing in a hurry when moving Bernard and hadn't then gone back to sort it out, and in all honesty it was completely useless. The piglets and their mums had lifted up the mesh all round the edge, and also pushed over several of the posts holding it together. Realizing there was little point in trying to fix the existing fence, we had the cunning plan of putting a new fence around the old one and making their area quite a bit bigger. This would hopefully reduce their urge to wander, and if we did the fence properly they wouldn't be able to anyway.

We duly got to work, and with the combination of my trusty post rammer and some thick mesh in an hour or so we had built the fence, including a new gateway, and they now had an area nearly twice the size it was originally. Of course while we'd been hard at work the pigs had all been off frolicking in the field and snuffling in the grass. When they really went at the grass they'd lift it out in huge clods and leave large bare areas of earth with piles of mud next to them. Sometimes I could fix these patches by putting the clods back in, but they were starting to do some serious damage to the field. We spent another thirty minutes or so chasing down the piglets, fervently hoping it was the last time. Once we had them back in and the gate closed, we slumped to the ground in exhaustion (actually I did, Alex always seems to have energy to spare).

The new fence worked! Well, it worked for about a month. We were starting to worry less about the pigs, and believed their troublesome days were behind us, until one morning we were once more greeted by piglets at the top of the field. It being a weekday it was a few days before we could do anything permanent to resolve the problem, though it was easier to get them back into their area now as they were more willing to follow the bucket. We'd also learned not to put down any food for the sheep first. Of course this meant that I was followed by the piglets and all the sheep, and on a couple of occasions the cows as well, all of them streaming behind me in a long line, almost in single file.

Unfortunately the piglets would sometimes break away from this pattern and run around, disrupting the sheep, who would then also start running all over the place. Then the cows would get spooked and start running around, kicking up their back legs with surprising agility and causing me some worry, both for my own safety and for the other animals. Muga, our Soay ram, didn't like this, and if it went on too long he would try to butt me to get me to put food down. Each morning I'd eventually get the piglets back in and all the food properly distributed, and each evening Alex would do the same. It was certainly helping us to get fit and adding some entertainment to our lives!

We patched up the new fencing. There'd been a couple of poorly joined areas which the pigs had found, and we even put a board across one area to limit their options. However, we'd learnt our lesson this time, and it was clear that they would work their way through the fence again unless we did something drastic, or clever! We therefore decided to try an electric fence. Several authorities maintained that electric fencing worked best with pigs, as they had very little hair or fleece to protect them from the jolt, and they would soon learn not to challenge it. As we still had the original electric fence we'd used with Bill and Ted it wasn't going to cost us too much to set up the new one. The only issue was getting the electricity out into the centre of the field, but in a stroke of near genius I decided to add an electric strand to the fence all the way: it would carry the charge and also fortify several more fences, including one in the orchard area which had been breached a few times!

Matt the shearer had suggested a style of electric fencing where instead of using the multi-stranded electric wire we used just a single strand of high tensile steel, and tension that between the fence poles; it was less likely to break and would still carry the charge well. That weekend we fenced away merrily, and in the end we had an electric wire along two sides of the orchard, much of the length of the small field, and into and around the pig and piglet area. Covered almost head to toe in mud (and more) I switched the new fencing on, and after cutting away a few bits of grass and leaf which were shorting it, I was happy that it was doing its job. Sure enough, we were soon rewarded with a squeal as one of the pigs hit the fence in the orchard, so we now knew it was definitely working, and they now knew they needed to avoid it.

The electric fence worked! Well, for about a month... The problem was that because the area was so muddy, the pigs would push up the mud into such a big pile that it ended up touching the wires, shorting them out, and then they'd only have the original fence between them and freedom again. Just to help matters they also ended up dragging the formerly electric steel wire half way round their area. We fixed it a couple of times, but in the end we gave up on that area and moved all the little ones in with Humphrey and the boys.

The problem was twofold: we couldn't tension the wire properly because the posts were so loose – and we couldn't ram them in any more without dismantling the mesh attached to them – and the mud was just beyond belief. The seemingly constant rain, combined with eleven pigs churning up the whole area with their sharp trotters and inquisitive noses, made it into a mud pit reminiscent of a war zone! Humphrey's area seemed secure enough and they all mixed well, all being big enough to avoid squishing, though Humph had to assert his dominance at meal times, on occasion by lifting one of the little ones into the air on his snout to a squeal of upset. However, fencing would continue to be a regular chore for us.

Gaffer Tape Saves the Day

Snowball's first litter had been born in the dark months of winter 2009 in the original hut in the orchard, one of two sets of piglets born that day. She had eleven piglets, one of which was dead on arrival (and which I buried in the other orchard area beside a tree). Finding a new-born litter of piglets is always a real wonder, and it was no different with Snowball. The day after they arrived I took her food into her pen, leaving it outside the front of her hut so she didn't have to move too far to get to it, when I noticed that one of her piglets was injured. It was probably the smallest of the litter, a tiny girl pig that fitted easily on my hand, and there was some blood on her side. When I picked her up I realized the injury was much worse than I thought: somehow half of the skin on one side had ripped off in a triangle revealing the flesh underneath (to this day I still haven't determined a satisfactory cause for the injury, and have never seen a repeat of any type).

To be honest I thought, 'Damn, we've lost another one, she'll be dead before the end of the day', but not being hard hearted enough to leave her as she was, or tough enough to kill her, I resolved to try and help her. I called Alex and she came to look at the poor thing. We thought the best thing to do would be to use our blue antibiotic spray ('for topical application to flesh wounds') to reduce the chances of infection, and then wrap her up. It was early in the morning and we wanted to get something sorted out fairly quickly, but which we felt confident would hold. We didn't have any bandages, but didn't know what to use, so looked around in the barn. We had plenty of gaffer tape as a byproduct of our ongoing barn conversion – and suddenly gaffer tape seemed a good idea as it's robust and sticks well to itself, but can be removed relatively easily.

I held the little piglet up and sprayed her side completely blue, managing to cover half my hand as well, which occasioned a comment or two that morning at work. Then while I held her still, Alex wrapped her up with the gaffer tape, making sure the skin went back on as closely as possible, and wrapping the tape round several times so it wouldn't come off. Despite all this attention the piglet made almost no noise at all – maybe she knew we were trying to help? We gingerly put her down and she gamely pushed towards Snowball who was lying on her side and allowing the others of the litter to suckle.

Despite all this effort I went to work assuming I'd hear from Alex that the little one had died, but not only did she not call, when I went home and fed them I saw that she was really getting in amongst the others and fighting for her place at her mother's teats. I still had my doubts that she would survive – given the size of the wound I was sure it would go septic – but when a couple of days later we carefully unwound the gaffer tape, which had survived its time in the pig hut in good shape, we saw that the skin was scabbing and perhaps even starting to heal. We managed not to tear it, sprayed it with antibiotic, and wrapped her round once again in some new gaffer tape.

As the weeks went on we continued to keep an eye on her, changing her tape and respraying her wound a couple more times, and amazingly after a month she no longer needed the tape. Piglets grow so fast, seeming to double in size every couple of weeks, that a wound which had taken up almost half her side was soon only a scar over her front leg. Even that disappeared really quickly, and after only another month or so the only way it was possible to tell that anything had happened to her was that she had a sharp pinch point in the white band around her body, and there is a clearly defined, sharp white triangle above her leg. Other than that, and a very slight stiffness in her leg, she is completely fine. We named her Gaffer, and Alex made me agree that we wouldn't take her to slaughter – so she became another permanent member of our growing pig herd...

WATER AND BASIC PLUMBING

One of the worst things that can happen to a pig is called salt poisoning, and it can very quickly kill it. This condition is caused by a lack of hydration, when in effect the pig becomes so thirsty that its body's water balance ends up going salty. Some of the warnings in the books indicate that when it's really hot this can happen within a day if the pigs can't get to water.

A standard rule for keeping any animals is to make sure they have ready access to fresh and clean water at all times. However, some animals are less bothered by a lack of water – sheep, for example, will satisfy the majority of their water needs from the grass (as long as it's not too dry) and won't drink that often. We have noticed with the sheep that when we're feeding them on concentrates they do tend to drink, but that's probably because there's very little water in the feed and they need the extra liquid to aid digestion. But all the authorities were agreed that pigs definitely need water at all times.

One of the skills we've picked up in converting the barn and managing the animals is basic plumbing. When all that is needed is to get water out to a drinking trough it's actually fairly simple, and a length of blue pipe, some joins and a tap or two, along with a trusty hacksaw, are all that is required to quickly add another 'drinker' to the system. Over the years I've learnt that we need to tighten and retighten some of the joins fairly regularly, and that if a pipe is exposed then it's likely the animals will knock it, so it's best to avoid exposed joins. For the majority of the animals maintaining their water supply doesn't take much more than just the occasional wander round to tighten a join by hand, or, far more rarely, with a pair of adjustable spanners. However, having said that, the rules with pigs are very different!

In part this is because of the type of drinkers I use, which are metal boxes around a foot and a bit square, and between half a foot and a foot deep (depending on the model). I used these with the OSBs and had no problems, and so used them with the Saddlebacks, assuming there would again be few problems. At first I was right: they are good sturdy drinkers, and seemed to hold up well. They come with a ballcock system already in place, and it's easy to connect them up to a length of copper pipe, which is then changed up to the plastic pipe. These connections present the problems, however, in that copper is fairly inflexible, so if the pigs move the drinker around violently, and if the blue pipe is fixed in some way, then the copper pipe will bend and either slow the water flow (OK), stop the water flow (bad) or break (very bad). This happened a few times before I learnt to keep the copper pipe to a minimum and let the plastic pipe do the work.

The next problem is the join between the copper and plastic pipes. This is the weakest point on the pipe, and on occasion the pigs will get themselves to a point of leverage where they can put maximum pressure on the join and it can then pull out, leaving the pipe happily spraying water around. Sometimes it seems they're in the mood to just bash the pipe around until something breaks!

It's not just the pigs being awkward which causes them to do this: there is a little more to it. Another problem for pigs during hot and sunny weather is that they can get sunburnt, especially on their light-coloured skin. For ours this is only really a problem for the strip around their shoulders, except for Bernard who is just about white all over and clearly has a much bigger problem. Their solution to this is to roll around in mud and cover themselves in a nice thick layer, which not only keeps them cool, it's also like putting on sunblock factor one million! But in order to have a mud bath in which to roll around they need some free-flowing water, and the best place for them to get this is from their drinkers, which is why they seem to attack them so regularly.

I'd like to say that we now have a foolproof system for dealing with the water which avoids all this extra pain... but sadly we aren't there yet. At any one time at least one of the drinkers will be leaking steadily, either from just a loose join or because of complete disconnection. This has two impacts on us: the first is that often the water pressure in the house will be rather poor, which is particularly irritating in the morning when I'm trying to shower and the water runs out in the middle of washing my hair and I end up standing under a slowly dripping shower covered in suds for five minutes while the pressure builds up again to have another go. The other is our water bill, which seems to get bigger every quarter!

One of the other joys about water is that it is one of the few substances which takes up a larger volume as a solid than it does as a liquid. What this means is that when water freezes, it expands, and this can cause taps and joins to split. While this is an irritating maintenance chore after the freeze when the water has thawed, it is a much bigger problem during the freeze, because if the water has frozen, then obviously it isn't getting to the animals, or at least it isn't getting there through the pipes! We therefore put a lot of effort into digging trenches across key parts of the fields so the water was delivered to each field with the minimum above-ground piping, which we insulated. This worked fairly well if the temperature only flirted briefly with -1 or $-2°C$ overnight, however if it was consistently $-4°$ or more, then we had problems!

During the heavy snowfalls of the winter of 2009/2010 all the outside piping froze, including the taps and all the water drinkers. The drinkers became, literally, blocks of ice, which was an added complication as this meant that even if we carried water out to them there was no point in trying to pour it in. We found our old rubber water troughs, and some other likely looking containers, and put these into the pig areas. Obviously these didn't have any water connected to them, so we had to carry the water out to them, one (or two) buckets at a time! Our mornings became much busier, going from around twenty minutes to feed everything, to forty-five or so to feed everything and carry all the water out. I worked on about twelve to fourteen water buckets in all, to supply the pigs' needs. Luckily the main water drinker in with the goats, cows and sheep was very big and initially didn't freeze solid all the way through, so at least to start with we poured boiling water into it, which usually managed to provide some water for them. However, when it got really cold we then had to fill that up with buckets as well.

Throughout the coldest parts of the winter this became our standard routine in the mornings and evenings. Part of the reason we had to refill the water containers every time we went out there was because the pigs insisted on flipping them over, sometimes while we were filling them – I think they enjoyed watching me slug back and forwards, carrying the heavy buckets through knee-deep snow. A little oddly I enjoyed it too: it made me feel a lot more in touch with the animals, and it was great exercise. Alex thought I was crazy as I'd carry the buckets out to the pigs and then jog back, only to do it again. She was far more sensible and walked back.

I also liked making little paths through the snow, and in particular the snow angel when I slipped over – though I pretended the angel was the reason I was on my back! The only thing which took a little of the shine off the exercise was when I occasionally managed to pour ice-cold water down my wellies. It only happened a few times, but it was just miserable as the water seeped through my two sets of socks and then seemed to freeze, turning my feet into lumps of frozen clay.

Some might argue that we were mollycoddling our animals and that there was plenty of water around – it was snow, after all – and I think there's an argument for that. However, to make sure that they were all right, and especially for the little ones, we felt that it was important to provide them with a fresh, unfrozen supply as much as was possible through the long cold days and nights. Even so, while I enjoyed the task, I have to admit to the huge relief when I could finally turn the water back on (after some tactical repairs), in part because it meant less effort, and in part because I then was confident that the animals, and especially the pigs, would have water the whole day through. The water supplies continue to test us, however, though I'm sure I'll find a solution soon.

Water was not the only challenging aspect we were facing: we were also concerned about the safety of our animals, particularly in regard to that enemy of all stock farmers (except cattle farmers I guess), the fox. We decided we needed something to help keep that risk at bay as well...

SHEEP, FOXES AND ALPACAS

The first winter after we bought the Mule ewes we were reviewing the sheep and hoping that our ram Muga had done his job and 'covered' all our ewes. Over the months we'd watched him pay careful attention to them – watched in the sense of seeing him and a ewe away in the distance: we certainly weren't getting too close or engaging in ovine voyeurism. When we took on the ewes from next door, Muga was nuzzling up to the first of them within five minutes, and a bit more than that five minutes later, so we were quite confident he was capable of the more practical aspects of breeding.

Muga's Responsibility

We had only two other worries. The first was whether he was covering every ewe in the flock, as some rams can apparently become fixated on only a few ewes, leaving the rest entirely alone. Given that we had three distinct types of ewe – the Soays, the Suffolks from next door and our Mules – there was a slight fear that he might prefer one of the breeds to the others, or some individuals across the breeds. In more organized and professional flocks the shepherd attaches a harness to the ram which carries a dye dispenser; this is positioned under his chest so that each time he covers a ewe he leaves a semi-permanent mark on her back. Truly organized shepherds will change the colour of the dye each week to get an idea of the likely delivery time of the lambs, and some even use different dyes for different rams if they have several running with the ewes at the same time.

We did none of this, however, leaving it to Muga to go his way without marking anyone, and just hoping we'd see results. It seemed to us a bit harsh to burden the poor ram with a harness, as they don't look at all comfortable, though we might just be fussing. Also, it wouldn't really have helped us a huge amount as I'm also not sure what we'd have done if he was ignoring some of the ewes. Perhaps arranged for a romantic night on the town for just the two (or more perhaps?) of them, some fine Chianti, and for dinner, the best sheep-mix pellets and grain available!

Rams are amongst the luckiest of male animals (if one uses the male-to-female ratio as a measure of luck, that is) in that they are usually kept at a ratio of one ram to between thirty-five and fifty ewes, and some flocks have ratios of 100 or even 150 to one. An average ram, as defined on the internet, can visit three to four ewes a day, so we knew that Muga should be able to get through our fifteen ewes every four or five days, all of which boded well for our lambing season.

The second worry was that while he might be versed in the practical side of the procedure, we had no proof that he could actually get the ewes pregnant. Bluntly, we didn't know if he was firing blanks or not as this was his first season as an adult, and we were concerned that the pressure to perform might have affected his confidence! But unfortunately we had no easy way to determine his efficacy, other than to wait patiently for the spring. This was a challenge, because the gestation period for lambs is about five months, so by my very rough reckoning, we should have been expecting to see lambs in March or April time. This was obviously a very rough guess, as we didn't know when or if he'd covered the ewes, or indeed if they were ready when he did. Ewes have a seventeen-day oestrus cycle, and it wasn't clear if he'd be sensitive to this cycle – though he should have been, because according to the books, the ewes would release a scent which would show him they were ready. Alternatively we supposed he might just go round on some arbitrary rota of his own creation.

A final consideration was that we'd introduced the Suffolks into the mix rather late, as they'd joined us in the first weeks of the New Year. However, we didn't know if they would still be fertile, or if their cycle would have finished. Most sheep breeds from the northern latitudes (and I'd guess the very southerly ones as well) have a fertility cycle which is dictated by the length of the day; thus as the days shorten going into winter the ewes will start their cycle, and will be ready to become pregnant. What we didn't know (and unsurprisingly wasn't clear from the literature), was whether this process stopped after midwinter? Did they get their Christmas presents and have their big feast and think, right that's it, no lambs for me this year...? With all this in mind we were starting to think about the spring and hoping for the arrival of some lambs – the first lambs to be actually born at Shaw Barn.

Foxes and Fox Control

At the same time one of our particular concerns, or at least one that we fixated on, was that we might lose lambs to foxes. I hate foxes. Not because they are cruel or kill for pleasure, because I don't believe they do – I think they just kill everything they can get their claws into, then drag the corpses away and bury them to provide a source of food for later. No, I hate them because they killed my chickens, several times. I'm an unforgiving person and I loved my chickens. I haven't written about it because it was a most upsetting episode – and a little repetitive, because we lost three rounds of chickens to foxes before we managed to get their orchard properly fox-proofed.

Foxes also like to take newborn lambs, and are especially likely to take them from ewes during their first lambing, when they're a bit unsure about what is going on and don't necessarily keep an eye on their little ones at all times. While most of our ewes had lambed at least once, they were still quite young, and the Soays had never lambed at all.

My neighbours had suffered an even worse time with foxes than we had, mostly because they had always had lots more poultry than we'd been able to take on. Unfortunately that made them much more attractive to the local fox population and they had lost many (several hundred I believe) chickens, and a whole flock of ducks just a few days after they bought them. After trying several forms of fencing and scares, Rob's next solution was to buy a rifle and shoot them . . . the foxes that is. Over the course of the next few weeks he went out every night with a spotlight on his quad bike to search for foxes. We didn't know this, and on the first night we were surprised and slightly perturbed to see lights wandering across Rob's fields! Fortunately he saw us and came over to explain, and panic was averted.

During the first week he shot about seven foxes, and believes he killed all of them. He killed several more as he went along, but unfortunately it made not a jot of difference. We later discovered that this was because foxes are territorial, so as soon as one fox was dead its neighbours would move in, and sadly they were all just as adept at getting to his chickens. After this he relied on electric fencing, but then realized that he was still paying a tax in chickens to the fox on a regular basis. Eventually he gave up on chickens completely. His last ones came to us, and ironically within three days they were all, bar one, killed by a fox.

After Rob's problems the whole fox issue was very much on my mind, and while out for a drink with a Geordie friend of mine we were discussing the challenge; he mentioned that many farms in the North East now used llamas to protect their flocks. Llamas are big scary animals happy to attack wolves, dogs and foxes, and if amongst sheep, a llama thinks it's a sheep and protects the flock (and the sheep seem to think it's a sheep too, and so are happy to follow it!). He'd heard that lamb losses due to foxes had dropped from quite a high percentage a year to none immediately after the introduction of one or two llamas into the flock. (He knew this because despite working in the Big Smoke he was very much into self-sufficiency and related topics.)

Under the influence of a couple of pints of ale I found the idea intriguing, and more importantly, still did the following morning when the alcohol had worn off – although I wasn't sure I liked the sound of the llama description: big, scary animals. While thinking about all this I was reading a passing article which mentioned alpacas, a smaller and less aggressive cousin of the llama, and the fact that they, too, would attack foxes (the down side was that they were vulnerable to big dogs and wolves, but neither of these are a big problem in Hampshire) – they sounded almost perfect!

This led me to investigate further on the internet, and add to our library of animal books, and much of what I read was very positive. I think there may have been some extra spin about their efficacy to defend against foxes, but it was corroborated in so many places it seemed likely that it was at least partially true. One of the stories on the internet was about a farmer who had quite a big problem with foxes, and had noticed that suddenly even more seemed to be going through his land; it seemed that his neighbour had bought some alpacas, and not only would the foxes not go into the field containing the alpacas, they also seemed to be avoiding the fields next to them. The farmer was shooting some of the foxes that were coming through his farm, but (per above) he realized this was only going to provide limited relief, and he was therefore considering acquiring some alpacas of his own!·

Researching Alpacas

As usual, the internet was amazingly useful in providing information (once again better than books), and I discovered quite a bit about alpacas... for one thing they were very expensive! Alpacas are one of those creatures that invite crazy speculative investments, rather like ostriches a decade or two ago, and prices were being inflated by people expecting to make a lot of money. This doesn't seem entirely sensible – after all, we'd been told you can't eat them (according to Alex my obsession with food value in animals is slightly perturbing...), you can't milk them, and they only produce one shearing of fleece a year, though it is lovely.

One major reason they were retaining their value was because they didn't multiply as fast as most domesticated animals. They generally have only one 'cria', or baby alpaca, at a time, and they have a long gestation period of nearly a year. The cria then takes two years to reach maturity. This compares favourably, in the sense of keeping prices up, to the classic 'boom and bust' cycle seen in pigs. Sows can have litters of around ten or more and are sexually mature in about six months, which can result in a glut and a subsequent price crash within two years, when two pigs could conceivably have turned into four hundred or more (though this is unlikely). People are encouraged to dive into pig farming when prices are high, which quickly results in overcapacity and a resultant bust. Alpacas would take far longer to achieve that kind of multiplication, and so the prices would be likely to stay high.

Further research on the web found us a few sites where one could view the wares of alpaca vendors, usually with prices. The least expensive females (always sold either pregnant or with a cria at foot) were around £1,500, though I saw a few up in the tens of thousands, and have heard tales of even higher prices for those with very pure colours. Males, on the other hand, as ever being the less useful gender, were down to £300 or so – not far off the cost of a cow. On this basis I came up with a cunning plan... I'd buy two males and use them as stud animals, and that would save us a lot of money. Alex and I then had a long discussion involving birds, bees and other things, and I was persuaded that this plan was not viable, so changed it to buying a single male to protect the flock (that being the original requirement), and potentially looking into the females at a later point.

We Purchase Alpacas

Back to the internet and I found someone selling male alpacas just a few miles down the road. I was finally getting used to the idea of geography! After a quick phone call, made by Alex as ever, we were off. Steve was a really nice guy, and showed us round and told us a lot about his alpacas. He had been one of the first to have alpacas in the UK, and had a very old white female who was quite lovely (though Alex thought she was a wizened, scrawny little creature...). One of the things which was immediately apparent when we went up to the males he'd selected for us was that they weren't actually that small, being about six feet tall and bigger in the body than a sheep. The other thing was how pretty they were: they have the loveliest eyes (especially when compared to the sometimes creepy eyes of our goats...) and huge eyelashes, and are easy to fall in love with!

When Steve had bought his property he'd had the fencing in the fields set up with additional channels between each field, which could be closed off to hold the alpacas in smaller pens. We really liked this idea and would come back to it later! He also told us that the alpacas rarely challenged the fences, and only if they were very frightened or very horny. He added that if we did get females it was important to keep them separated from the males by a decent distance when they were in season, both to prevent fighting amongst the males and to avoid an unplanned cross. Colour purity is considered rather important by many alpaca keepers, as it makes the fleece more valuable and easier to handle. Avoiding an unwanted cross was even more important when one considered that this would mean that the female would be busy for a full year before there would next be a chance to breed from her.

As we were intending to actually buy an alpaca, Steve was quick to inform us that they (and llamas) are very social animals, and he insisted that we would have to keep a minimum of two of them to prevent them pining and dying of loneliness. I'd read some mention of them being very social in books, but it hadn't been that clear. We knew that pigs were social creatures and could die if left alone, so it didn't seem to be unreasonable for the same to be true of alpacas. This meant we had to buy two ... and suddenly our lamb protection scheme didn't seem to be so cheap; however, we were so far down the mental path to alpaca ownership that this could not stop us! We told ourselves that the alpacas would live for quite a while, twenty years, and over that time would more than pay for themselves, especially if we sold their fleeces. Steve clearly believed that we needed to have a minimum of two alpacas to keep them happy, and it wasn't that he was just an excellent salesman.

We chose two alpacas, a dark brown one with a white front (which Alex named Algenon, or Algy for short) and one that was a tan colour all over (we named him Verdigris, though I can't really explain why – maybe because it's a nice word, and perhaps because he was almost – at a stretch – copper coloured). I believe Steve said something about these two hanging around together more than the others, and in passing something about Verdigris being a bit more aggressive, but we carried on regardless.

Getting them into the trailer was quite a chore. They were already rather spooked by the arrival of two strangers, they were fairly big and strong, and they were not hugely interested either in food, or in going into the dark box of the trailer; but we managed it eventually with a little push and pull, and then happily headed home with our new lamb guards!

We released the two of them into the field and they started investigating their new home, and their new cohabitees. I think the sheep initially intrigued them – they certainly seemed to peer at them closely, perhaps wondering why their necks were so short! They cordially ignored the cows, and were cordially ignored in return. They got a bit too close to the goats, who rammed them in a rather desultory fashion. The alpacas were clearly a little nervous and backed off quickly, but now knew to keep away from the goats. This all happened rather quickly, but within a week Algy and Verdigris would run with the sheep at feeding time, though they'd let the sheep go to the food first, in part from a natural desire to avoid being crushed by the stampede, and in part from their continued shyness.

The Alpacas Settle In

An important, and perhaps over-stressed note in one of the books informed us that alpacas don't preen each other in the wild and therefore really don't like being stroked or touched, which, given the lovely softness of their fleece, seems a bit of a travesty! This meant that even when they did calm down and allow us to get close, we would have to be careful not to be too tactile with them – unlike, for example, the cows, which Alex likes hugging around the neck if they are willing to stand still for that long.

After several weeks Algy, who was clearly the more confident of the two, would occasionally eat from my hand, although he was still very tentative: at first he'd take just a nibble or two, and then he'd back off, before coming back again for more, very much like our original two Soays, in fact. Verdigris, however, would not come close enough to eat from my hand. Whenever he was *almost* close enough he would spook, possibly because all the sheep would mill around at feeding time, and possibly because White Face or Muga would often try and get to whatever food was in my hand, pushing out anything else which dared to get in the way!

Alpacas are extremely fastidious creatures, especially about where they go to the toilet. Sheep, goats and cows will go wherever they are, with no thought as to location, or indeed where any other creatures may be – I've seen a sheep leap out of the way of a flood of urine coming from Wrath. Alpacas, by contrast, tend to pick a spot and go to it religiously until at some arbitrary point they decide it is full, when they will find another spot. They also both go in the same spot – often one stands near while the other one is going, waits for them to finish, and then uses the same spot. Actually they seem to have several spots, but they really do only go to those spots they've designated.

After a week or so of use each spot turns into a circle about three feet in diameter and stays that way, never getting any bigger, until they abandon it, when it is slowly reconsumed by the pasture. In Peru, where they are used as pack animals, this is a real boon because it means they don't make a mess on the roads; indeed, their pack handlers keep special bundles containing the results of previous visits, which they roll out in a mat at the side of the road, and the alpacas use those, keeping everything clean and controlled!

A well-known gardening radio show once mentioned that alpaca manure was supposed to be one of the best types available, although they did admit that this information had been provided by an alpaca keeper who might not have been entirely unbiased. One thing I would say is that the grass is always extremely lush in the circle immediately around one of their toilets, and it does seem to grow back fairly quickly once they abandon it.

So in a matter of a week or two Algy and Verdigris knew where everything was, how to get extra feed (as well as the grass obviously), and where their toilets were. Our lamb protection system was therefore in place and ready for its first test!

THE LAMBING SEASON

Most shepherds like to have a relatively short lambing season, and they also like to know reasonably accurately when the lambs are due. In the main, this is so they are in a better position to be able to help their ewes if they get in trouble, particularly if it is a large flock. The vast majority of ewes have little trouble in lambing, but some do, and it can be fatal, both for the lambs and for the ewes. In countries such as New Zealand with super-sized flocks they tend to just let the ewes get on with it, and if they lose some at lambing that is sad, but they take the view that those ewes or lambs were weak, so overall it is a good result, because it keeps the flock strong.

In the UK the margins are so thin that every lamb counts, so many farmers, and vets, work through night after night to ensure that all their lambs get the best possible chance of life. Knowing all this you might think that we'd be somewhat stressed about the timing of the lambing cycle, and that we'd be ready to keep a regular nightly watch. However, we weren't worried, because – somewhat surprisingly for us! – we'd thought this through in advance.

Our secret was the use of Muga as a ram. Even as a full grown ram he was two-thirds the size of the Mules or the Suffolks, and so any lambs he sired would also be small. This, we hoped, would mean that any complications with those ewes would be far less likely. We'd been reliably informed that Soays would sort themselves out by both sets of people we'd bought them from. We have since learnt that they do occasionally have problems, but this is much rarer than in the commercial breeds.

I read up on all that we might need to do, and made sure we had the basics to look after the lambs and ewes – though even the list of basics I put together is debatable. One of our books had a list of around twenty items which would be needed in order to be prepared for lambs, ranging from old towels (check), through antiseptic spray (check) to elbow-length latex gloves (well, we're not really like that), and spare colostrum (no, not as yet). Colostrum is the ewe's first milk and is full of antibodies to protect the lamb until its own immune system gets going. It needs to be fed to young lambs in the first twenty-four hours of their lives, and the mother will normally provide it. We'd only need spare if we lost a ewe during the birth, or she didn't have any milk, and we had the Magic Shop down the road or the vet on the phone if that happened. Ditto for the shoulder-length gloves, though a few people (including a vet) have said that it's much better not to use gloves because when you're in the ewe you'll be able to feel what's going on better, and to just wash your hand and arm thoroughly afterwards. I rather hoped we would not have to find out.

However, if we had a ewe in trouble we had a fourfold back-up plan: we would call John the farmer, who'd sold us the Mules originally. If he was unavailable or couldn't help, then we'd call a friend of Alex's who was a vet, and while she was too far away to actually come and visit, we hoped she would be able to help us over the phone. In parallel to these two we'd speak to our neighbour Claire, who went on a lambing course and hopefully would be able to help us. Our final back-up was to call our vets, who were only fifteen minutes away and would hopefully be able to come in time.

We felt we were ready for it, we just didn't know when it would happen! So we continued to feed them all twice a day, checking the water, especially when it was frosty, and keeping an eye out for lambs. Then on 7 March at around eight o'clock, I opened the curtains of our bedroom and stared out across the field. All was well with the world, and I was smiling benignly at the part of our realm I could see (almost everything visible from that window is ours up to the distant tree line, and it still feels amazing to survey it all) when I noticed one of the sheep in the distance right next to the woods with a couple of blobs of white next to her. I let forth an oath of surprise and pleasure, threw on my clothes, grabbed some feed and ran out to see the sheep.

After feeding the main lot, who were uninterested in the potential for lambs and would have plagued me mercilessly had I not seen to them first, I made my way quickly, and then more carefully, towards the ewe and potential lambs. It was Horny, and it seemed she had had three lambs, which was just breathtakingly amazing. The wonder of new lambs, looking so cute, is almost impossible to understand unless you've seen it yourself – even more so as they tentatively totter around, perhaps knocking into each other, or their mother, and curiously meandering towards me. They were mainly white but had darker splodges on them, clearly something they'd inherited from their father, perhaps the genes which might also have produced a piebald in another Soay. Horny had done all the hard work and successfully delivered her lambs, and there was really nothing for me to do. I was expecting a nurse to pop up and tell me to go and boil water or something equivalent, and in the end all I could do was head back into the house to find a camera to take a picture of all these lovely lambs. It was about this time that Bob Marley singing 'We be lammin'' started running through my head...

I have a policy on counting lambs. The majority of lamb fatalities occur within the first twenty-four hours – in other words, if they last a day they're probably going to last the whole innings. So I never count a lamb as part of the flock until it's been alive for at least a day, partly through superstition and partly through a sort of distancing mechanism, to try and reduce the upset of losing a lamb. Fortunately all Horny's lambs were fine the next day, and I could rejoice, officially add them to the flock, and relax.

For the next three days she stayed with them, and wouldn't join in with the regular feeding, so I'd take her several handfuls of feed. She would need extra feed as she had a triplet. Many shepherds would have tried to mother one of the triplets on to a ewe that had had only one, to balance out the load on each ewe and reduce the risk of losing one of the triplets, or even worse, the mothering ewe. We didn't really have that option, not only because none of the other ewes had lambed, but also because this tends to be an operation best carried out in a barn, or some other restricted space, and we did not have the facilities to support that. To mitigate the risk I fed her extra, and always try to with those that have had triplets, though given that they also tend to be extra hungry and therefore more pushy they often seem to sort themselves out anyway!

After Horny had lambed, each day we'd wake up hopeful that more lambs would have been born. Each day we'd count the ewes as they came up for feeding, hoping there would be one missing from the food line, because this should mean she was somewhere out in the field, hopefully looking after her newborn offspring. There was a gap of a week or so, and then two more ewes dropped their lambs: White Face, who also had triplets, much to my pleasure, and another Mule which had twins! Having both lamb on the same day was a special treat, as we had five more lambs to delight in. We were immensely proud of them, strutting around almost as if *we* were the ones who'd done all the hard work!

We were also particularly excited that our lambing average was above 200 per cent – that is, we were getting more than twins per ewe. This is a key measure for shepherds, with 200 per cent being the upper target for most breeds of sheep. While we knew it would inevitably drop below that target, it was exciting at the time! The remaining two Mules lambed over the next few days, having twins and a single, which still left us with a great average, and great delight.

One of the duties of a shepherd is to dock the tails of new lambs and to castrate the boys. I had read up on the options, and selected the one which seemed to be the least trouble for both parties, and which didn't involve blood (or a vet!). It required a device called an elastrator, and involved getting an orange rubber band about the size of a polo mint stretched over and then released on to the tail, and another round the base of the scrotum. I have to admit to being rather intimidated by both the very functional device – it consists of four prongs upon which the band is rolled, before the levers are squeezed causing the prongs to pull apart, thereby opening up the band – and the act of castration itself. This is not something any male does without at least a little sense of vulnerability!

However, my trepidation meant I was late, because all castrations by this means should be done within the first seven days of the birth to minimize the discomfort (cringe) for the lamb. By this measure Horny's lambs were all too old – not that I could have caught them, even had I wanted to castrate them.

Fortunately that weekend Matt was due to shear our goats, and he showed me how to work the elastrator, and also his no-nonsense manner made me grow a bit of backbone! He did admit that he, too, wasn't comfortable when doing the deed, and that his wife gained some amusement from that fact. That day we 'did' the lambs that we could catch and which were young enough, which fortunately was the same group. The recommendation from the books is that one castrates and docks lambs in their first two or three days of life because that's when they're still young enough to be caught, and on this one point at least the books all agreed, and were all absolutely right. From then on my ritual with newborn lambs was not just to take photos, but after their first full day of life, catch them, dock them and, where necessary, castrate them.

All the Mules by this time had had their lambs, but surprisingly none of the Soays. We knew that the Suffolks, if pregnant at all, were still a couple of months away, but the Soays really should have no excuse. Another month later and they *still* hadn't dropped, and we were getting more and more concerned, wondering if they were in fact pregnant, or if Muga had failed us, or if we had some other problem and they had miscarried or something similar.

In addition we were now getting near to the time when the Suffolks might drop as well. Matt came to shear our sheep in May of that year, and he refused, politely, to do the Suffolks on the basis that when they're being shorn they can be turned around a bit, and he'd had an experience a few years before where a set of pregnant ewes he'd shorn, lambed a couple of weeks later and all seemed to have problems. So for our own good he said he'd rather wait and do them later in the season, they'd just be a bit hot for a bit longer, and we'd have to keep an extra eye out for fly strike. Two weeks later and all the Suffolks lambed within three days of each other with no problems at all. This still left the Soays, and we'd almost given up hope.

Trying to maintain optimism, we decided that at least some of the Soays had to be pregnant as they were looking a trifle fat, but it was difficult to be totally sure. Still, we held out hope – and if they weren't pregnant we would somehow have to put them on a diet and cut their grass rations! We knew Soays always like to be a bit different. This was especially obvious in their independent ways, but was further demonstrated when they finally decided to lamb. They had a particular MO for lambing. The first thing I would notice was that a little brown ewe would be missing, though I would have to count several times just to be sure. (During this period I was monopolizing the feeding of the animals, much to Alex's chagrin, but as she was exercising a lot at this point the only way she'd have been able to do both is to get up another thirty minutes earlier, at five o'clock, and she wasn't willing to do that just yet!) Then I'd search the fields for the ewe. This, one might think, would be easy: brown ewe, green grass, should be obvious. Sadly this is not the case, because when a Soay is sitting down, as she most likely would be, she will look like a mole hill or a smudge of mud from any reasonable distance, so I needed to trudge nearly the whole length of the fields to have a chance of finding her.

The first time this happened I could not see the missing ewe anywhere. I searched up and down the fields, looking everywhere, and even went back to the feeding area to re-count the remaining ewes. There was definitely one missing – and suddenly my heart was in my throat, what if she'd got out? Or been taken away by a fox in her hour of weakness? So back I went, looking for her, with a paper bag half filled with feed held tightly in one hand.

Then as I was passing through the tree line between our main field and our meadow which runs down to the stream, my eye caught a flash of movement – and there she was, standing next to one of the trees with a little lamb at her feet. She was so well camouflaged that if she hadn't moved at just the right moment I might not have seen her. The first to lamb was Mouton, and she seemed very proud of her little one, and not at all interested in me getting too close. I threw her some food which she started to eat, but if I got too close she backed off through the trees and nettles. Fortunately the next day I was able to get close enough to catch her lamb to dock its tail, though luckily no castration was required as it was a girl!

That first day I was late getting into work after all that searching round the fields, and to avoid this I went out extra early for the next few days. Fortunately I now knew where to look, and found them much more easily as they all seemed to like to retreat into the cover of the tree line.

After a week all but one had lambed, and it seemed likely that she wasn't pregnant as she looked as thin as the other ewes did after their lambing – and indeed she didn't lamb that year. It's possible that the Soays were so late because they were all young, perhaps being too young when Muga was first going round the Mules, but old enough when he was re-energized by the Suffolks!

Real farmers or shepherds are surprised and amused by the length of our lambing season, as compared to their endeavours to ensure the season is as short as possible. One of the techniques involves keeping the ram away from the ewes for two weeks just before the peak of their season, and then releasing him into the field. This can, apparently, all things being equal, result in a lambing season of just a couple of weeks. From the first lamb at the beginning of March, to our last lamb born in mid-June, our lambing season was all of three and a half months – though we would say that this was no bad thing, really, as we extended the joy of the lambing season to its maximum extent, and given that a newborn lamb is probably the greatest reward from keeping sheep, we were not unhappy with the outcome!

We had a total of twenty-three lambs that year, of which seven were pure Soays. In the end none of the lambs or ewes required any intervention, and only one was lost, a lovely little brown lamb, one of twins. I saw him the day he was born, but the next day he was gone and I could find no trace of him, and it left me wondering if a dog or fox had got the poor little creature. Still, twenty-three was fairly astonishing, and more than we thought we'd ever have.

Success of the Lamb Protection System

Through all this Algy and Verdigris seemed to be doing their job, apart from the one lamb that disappeared – which, in their defence, could have been a number of things. They might not be patrolling all the boundaries on a rigid schedule, but they generally seemed to be sticking with the flock, and they showed extra care for the ewes with newborn lambs. Broadly, however, we were pleased with our successful deployment of the lamb protection system . . . until Matt our shearer told us that he'd not lost any lambs and he didn't have any alpacas. He had also heard that if you had more than one alpaca then they would form a flock themselves and not protect the sheep! We are not totally convinced of that, however, because certainly with one of the ewes the alpacas sat about six feet away from her for several hours after she'd just given birth to twins, and it seemed clear (to us at least!) that they were looking after her.

So we sang the praises of the alpacas to our neighbours Rob and Claire. Rob was frustrated with the lack of impact of his fox culling, and Claire thought the alpacas were lovely, so they decided to go and see Steve about perhaps getting a couple of their own. Rob came back a few hours later laughing: Steve had been telling them how gentle the alpacas were when one of them got a bit frisky and kicked him in the family jewels . . . apparently Steve had carried on manfully, trying to pretend he wasn't in pain and that they really weren't that violent!

Suffice it to say, they didn't buy any. Our own two have never been violent with us, ever, possibly as we very rarely challenge them in any way. Nevertheless we're also always careful around them, just in case, as it's never sensible to take their good nature for granted (or that of any animal).

The Alpacas Need Shearing

Unfortunately, Matt, who shears our goats and sheep, does not shear alpacas as he says they're more trouble than they're worth, so we needed to find another shearer. I had been given a contact number of a Kiwi shearer who comes over each summer and does as many alpacas as he can, but we weren't able to organize a mutually convenient date. Then Rob from next door reminded us (to be fair he must have told us a dozen times!) that the sister of the man doing our hay, whose name was Suzy, was a shearer who did alpacas, so Alex phoned her and organized a time and date.

At this point we'd never caught the alpacas, and I for one was a little nervous – I was even wondering if I should buy a cricket box! However, Alex had asked Suzy how we should catch the alpacas, and she told us that if we had a line of rope held at around chest height the alpacas would not challenge it and we should be able to manoeuvre them into a stall made from sheep hurdles. This we duly did, and it was astonishingly easy as they really would not challenge the rope.

When Suzy arrived she brought out a canvas mat and some metal stakes which she hammered into the ground each side of the mat, about eight feet apart in total. I was somewhat perturbed by these preparations but as she was a champion shearer – she enters competitions and all sorts – I certainly wasn't going to argue. She then directed us in how to catch one of the alpacas, and looped ropes around his front and back legs and tied them to the stakes; we then manhandled him to the ground, and she tightened the ropes to stretch him out so he couldn't move. She also put a sock over his mouth, which seemed cruel, but she told us it was to stop him spitting. Well, it wouldn't stop him actually spitting, but it would catch the spit when he did, rather than it spraying all over us and her. She then very efficiently sheared him, with the fleece coming off in chunks which would fall into separate fibres in our hands, very unlike sheep or goat wool.

While we were in the process of getting him up again the sock came off his mouth and some of the spit came out. It was partially chewed grass and bright green, but combined with his saliva or stomach acids or both, it was absolutely foul, and Suzy told us that once it was on clothes it was impossible to get out! We were extra careful around him and fortunately he chose not to spit at us. Suzy also trimmed his hooves and sanded down his fighting teeth as these are sharp and can grow long, and could cause real damage if the alpacas were to fight each other.

When we released him after these ministrations he looked comical he was so very thin and ill proportioned, especially his overlong neck. A similar process followed with Verdigris, who was just as unhappy as Algy with the whole thing! Unlike the goats, they did seem able to identify each other after they were shorn, and didn't fight each other for dominance in the way the goats had.

The shearing left us with two bags of lovely alpaca fleece, which we had no idea what to do with . . . so it was back to the internet for me. I looked at eBay and established that alpaca wool was sold in small packets for a fairly substantial amount. However, I figured that separating it all out and bagging it was more effort than we were willing to invest at that point, so I weighed both bags and put them on eBay as whole fleeces, thinking that this would cause some crazed bidding and we'd get a decent price for them... But in the end we only had one bidder for both bags, and she paid us about half the cost of the shearing! Still, she was completely blown away by the amount of fleece she was getting, and so at least she was happy!

After catching the alpacas so easily our confidence soared, and I was feeling happier about keeping them. For several months I'd been wondering if they were really worth the effort, and if they really had helped with scaring off the foxes. I wasn't sure, and it had seemed impossible to manage them. But now I was full of confidence again, and a few weeks later decided to use the rope technique to manoeuvre them; I wanted to separate the sheep from the alpacas as part of a plan to get some of them ready for 'holiday'. Alex was out, so I had to use ingenuity to achieve my objective: I decided I'd put a rope across the top of one of the gates, and then herd the sheep through underneath it. The alpacas wouldn't challenge the rope and the task would be completed and there would be much back-patting all round.

Having set this up, I started herding the sheep towards the gate, the alpacas of course coming along too. My herding method was to half fill a bucket with feed and shake it so that the rattling sound would attract them. I guess in reality I was leading the sheep, not herding them – perhaps I was a sheepleader instead of a shepherd? The sheep duly started to pass through the gate without issue. Pouring out the feed in a spot to keep them in their new area, I headed back to the gate to close it as soon as the last sheep were through. The alpacas were by now approaching the rope line, and without pausing they gracefully ducked their heads and went straight under the rope!

I was so surprised I literally stood rooted to the spot for a few minutes. It seemed they would challenge a rope when shown how, and now I had them in the wrong field as well!

Through the use of more food I persuaded all the sheep and the alpacas back into the first field and reconsidered my strategy. In the end I trapped the sheep in a set of hurdles and walked them through individually, which was far more effort! There was also the niggling thought that the alpacas now knew how to deal with a rope, and the next time we needed to shear them, Algy and Verdigris would easily duck our efforts ... but that was a concern for another year!

The alpacas were generally easy to care for, though they had a few foibles. Occasionally they would fight and we'd hear their weird, high-pitched warbling as they chased after each other, but they didn't seem to hurt themselves and they soon subsided, so we worried not. Then one day we heard the usual warbling, but it seemed to go on for a while, and was then followed by a strange, almost sighing sound; so we popped round to have a look, and were amazed at the sight of one of the alpacas apparently sitting on one of the ewes.

After some consideration we realized two things: the first was that the sound was coming from Algy, and the second was that he was trying to 'cover' one of the Suffolk-cross ewes. Her eyes were going wild, but she couldn't move as he had her pinned... Unusually Algy allowed me to get very close to him and seemed uninterested in moving, and I didn't feel it was entirely appropriate to try to push him off (or indeed to keep watching!). Some quick research in our books and on the internet provided some information.

Apparently when kept with sheep, male alpacas have been known to become amorously involved with some ewes. This is not a particular problem, though it does tend to bemuse and upset the ewes, especially as alpacas usually take around twenty minutes to have sex, as opposed to the lightning few seconds of a normal ram! Unfortunately, or thankfully depending on how you look at these things, there is also no chance of sheepacas or alpeep, as they are from such different families (alpacas are part of the camel family). After Algy's allotted time he arose and wandered off (no cuddling it seems) and she got up and wandered off as well, to all intents and purposes no worse for wear.

Algy has 'jumped' a few other ewes since, always a Suffolk (and possibly the same one), but we've never seen Verdigris at it. He is either shy or more discreet!

Consolidating our Assets

We now had quite a variety of animals and decided to concentrate on looking after them as best we could, and didn't bring any more on to the smallholding for quite a while. We even had a chance to start reducing them a little...

MANAGING THE PIG ENTERPRISE

After Gaffer's early brush with mortal danger we kept a special eye on her, made easier by her distinctive triangular marking. But once she'd recovered she grew faster than her peers, and having started as one of the smallest she was soon one of the bigger ones. Her stubborn and persevering nature meant she was particularly good at getting the maximum amount of food, even (or perhaps especially) if it meant pushing one of her litter mates out of the way. As per my agreement that we would not take her 'on holiday' or sell her, we knew she would stay with us even when all the others of her generation had gone.

Sorting Out Numbers

At this point we had over forty pigs and were getting through feed at the alarming and expensive rate of a ton every two weeks or so. We needed to reduce our numbers. We could either wait until the younger ones were grown and take them on holiday, or sell them on to someone else to finish them off. If we went for the first option we'd be buying tons of food for several more months, and then have tons of pork, probably necessitating more freezers. If we'd been confident we could sell all the pork we might have tried it, but it wasn't something we had much experience with, and felt it was rather too risky. At this point the second option was definitely preferred.

It was our first attempt at selling livestock and we didn't do a particularly good job of advertising them. Over a couple of weeks we only managed a couple of notices in the Magic Shop, as we were both very busy with work at the time. We didn't have many takers and eventually sold most of them to a local farmer who specialized in finishing pigs. The biggest cost with pigs is feed, which is why most farmers who specialize in producing weaners like to sell them as soon as possible – six to eight weeks is considered the most reasonable time. Our 'weaners' were more like four months old, and so we should really have been able to get a bit more for them. However by this point we were fairly desperate to reduce our feed costs, so we sold them to the farmer for standard weaner prices. I think he got an excellent bargain, but we were just glad to get the numbers down to something manageable. We did keep three of the oldest ones ourselves to fatten them up for a few more months before sending them on holiday; we kept boars on the basis that at least Humphrey couldn't get them pregnant if he got in with them!

The plan with Gaffer was fairly basic, I have to admit. We would let her grow on until she was the same size as the adult pigs, in effect until she was around twelve to fifteen months old, and then decide whether to introduce her to Humphrey, or another boar, or just keep her as a pet. On the basis that Humphrey was her father and grandfather we were not too keen on the introduction, but on the other hand I felt it was important that she somehow earned her bread and butter.

We were getting much better at rotating the sows in with Sir Humphrey. We wanted to ensure he had constant companionship, as this seemed to reduce his wanderlust, though it meant we always had another litter on the way. Hacker had been in with him first, for three months and then given birth to another litter. Next we put Malcolm in with him. Three months later she was very likely pregnant, so we swapped her out and provided Bernard as his new companion.

The problem now was that we were running out of areas to keep all these pigs separate, even though we'd managed to sell off a whole lot. We needed at least five areas to keep them separate:

* Sir Humphrey and Bernard
* Malcolm (probably pregnant)
* Hacker and the piglets
* Maturing boars
* Snowball and other unnamed adult sow
* Gaffer

We knew we could put Gaffer in with Snowball and friend, Malcolm or potentially Hacker, and in the end we decided on Malcolm, as we figured it was less competition for Gaffer, and when Malcolm did give birth it was unlikely that Gaffer would be a threat to them. Malcolm was quite a bit bigger than Gaffer, perhaps four times her volume all told. As is the way with pigs, at first there was a little pushing and scrabbling at each other, but Gaffer realized she was outgunned and tended to back off quickly. The most obvious time for fighting was over food, and we were careful to ensure we put out at least two piles so they could eat separately. After a week or so they were comfortable with each other and there was only the smallest amount of natural pushing and shoving. We'd reached a new state of happy equilibrium, and all was well for a while.

More Problems for Gaffer

Then one day while feeding them we noticed that Gaffer had a bit of a limp. Actually when we watched her carefully we realized she wasn't putting any weight down on her back right leg. We checked her leg carefully and couldn't find any cuts or lumps on it, and the bones seemed fine, but she would not put weight on the leg. She was still eating fine, and getting herself around relatively well. We immediately moved Malcolm out of the area as she showed little sympathy for Gaffer's injury, pushing her out of the way with as much casual force as usual. Gaffer still shared a fence with Hacker, her piglets, and now Malcolm, so we figured she shouldn't get lonely.

Gaffer was one of the pigs we felt most attached to, and Alex was particularly keen that we call the vet out. Thinking that we should get her checked early on in case she went downhill quickly, I agreed. The vet couldn't find anything wrong with her either, which made us feel better because it meant we'd checked all we could, but also worse, because it didn't explain what was wrong with her. The vet gave her an injection of anti-inflammatory drugs, and said that if the leg didn't get better in a few days we could pick up another injection if we were happy to give it ourselves (we were, we knew the bucket trick!). If that didn't work then he said we should consider putting her down.

This somewhat upset Alex, and it felt to me a little premature as well. We knew Gaffer had great spirit and we believed that she'd recover well if given the chance. In one final parting shot the vet mentioned that he thought Gaffer looked fat! We were all three offended by that statement. Gaffer was a fine-looking example of a pig, and Alex and I spent several minutes assuring her that the vet didn't know what he was talking about.

Gaffer's leg did not improve over the next three days, though it didn't get any worse. Despite some doubt over the efficacy of the first injection, we decided to go for the second one. Gaffer really did not like the bucket, and showed surprising agility in her efforts to avoid it, almost belying the limp! She also had amazing perception in quickly spotting me sneaking up on her while she was eating – normally the pigs are completely oblivious to anything else when there's food to be had. But once the bucket was on her head she calmed down, much as we had expected, based on previous experience. We were worried that all the running around might have stressed her leg further, but there was little choice if we wanted to help her and get the injection into her.

I'm not sure if it was the injections, or maybe the workout, but after this dose of anti-inflammatories she did seem to improve. Slowly her limp diminished, until after a couple of weeks it was almost gone completely. This reinforced our trust in her perseverance and spirit.

Around this time it was our wedding anniversary, and we decided to go away for the weekend – though by now, going away wasn't just a 'pack the bags and lock the house' kind of decision. Fortunately Alex's mum, Sue, kindly agreed to look after the animals while we were away. We had an excellent break, and after two refreshing days returned to the barn, and during that evening's feed performed our usual check of the animals, not expecting much as we hadn't been away for long. However, there was a surprise for us: Gaffer – not that she was limping, but something else altogether.

Motherhood for Gaffer

I called Alex over and said, 'That is a pregnant pig!' pointing to Gaffer. She had all the obvious signs, all her teats were swollen and her milk bags seemed full and tight to the touch. We couldn't get any milk out of the teats, but that doesn't usually happen until very close to actual farrowing. Alex thought it might be a false pregnancy or something else of that nature, because as far as we knew there was no way she could be pregnant. We had ensured that she never spent any time with Humphrey, apart from when she was very small, before she was even weaned. Nevertheless I was still convinced she was pregnant, by what mechanism I knew not, and I thought she'd have her litter within the next twenty-four hours. This is a bit of a joke between Alex and I, in that once their teats are swollen I always predict a sow will farrow within a day, and I've been wrong by as much as a week before.

This time, however, I was completely correct, and the next morning we came out to find Gaffer with just a single solitary piglet. We wondered if that was the first of the litter and we'd interrupted her farrowing, so we left her alone. Several hours later we checked on her again to see how she was doing and if she'd had any more, but there was still just the single piglet. We looked through the straw of her hut to see if there were any others avoiding us or any stillborns, which the sow will often cover up and hide, but despite a thorough search we couldn't find any more at all.

This was good, because we really weren't ready to have many more pigs again at this point. Given the size she'd been we were amazed that she'd only had the one, almost as much as at the fact that she'd been pregnant at all! Still, the lucky piglet had access to a surfeit of milk, and we knew she would grow fast and well without the competition that siblings would have provided.

There was still the question as to exactly how she'd got pregnant, and indeed, who the daddy was. After some thought we decided that the father was probably one of her half-brothers (all of which were long gone at this point), and part of the reason for the small litter was the very young ages of both sow and boar. For us this was yet another lesson learned: we'd have to split up the boys and girls into separate groups when we weaned them, to avoid any other unplanned teenage pregnancies.

Gaffer seemed fine, though with a new piglet, her teats huge and swollen, and her leg not completely healed, she wasn't the most comfortable of pigs. The arrival of the piglet did not seem to slow her healing leg, however, and it wasn't long before she was walking normally – it was as if she'd never had a

The Irreplaceable Value of Baler Twine

A brief aside on baler twine. When hay or straw is baled, at least in small bales, it is held together with baler twine. In our experience this is generally bright orange and will be in two loops at a third and two-thirds of the way along the bale. When first I started dealing with hay and straw this was at best irrelevant to me, and at worst a nuisance. In the early days before I'd learnt to carry a penknife whenever dealing with the outside, it meant squashing the hay or straw awkwardly out from the twine, and then throwing the twine away. It didn't take long for there to be piles of the stuff.

Worse than this, the twine commonly in use really isn't very environmentally friendly as it is generally made of a plastic type of material. This helps it to be nice and strong, but it doesn't biodegrade particularly quickly (though it is allegedly designed to do so). With my generally green thinking, this seemed like a bad thing, and for a while I was wondering what we should do to get rid of it.

This all changed when I realized that it was actually the single most useful item to carry on one's person at all times, even more so than a penknife. In fact once it is combined with a penknife it becomes an almost unbeatable source of innovative bodging and fixing around the smallholding. We have used baler twine to variously fix fences, close gates, fix together hurdles, affix trees to posts, string up CDs to scare away crows, hold open doors, lock doors, keep tarpaulins down, put up 'Close the Gate' signs, affix electric fence posts to gate posts, tie bird mesh together and to the fences, hold water pipes to beams, build temporary shelters, fix Red Mason bees nests in place, and last but not least, to bale hay! For a while some of our fences seemed to be made more of baler twine than steel mesh. In fact it's so useful that Alex has taken to tying it randomly to fences around the place so that if we need it we don't have far to go to get it. I tend to keep several lengths in my pockets throughout most of the year.

I suspect some day, probably too soon, some official body will insist that all baler twine degrades within a year (after all, the hay or straw usually does), and then we'll be left without its magical properties and will have to find something else. But whatever the replacement is, it will need to be pretty special as I sometimes think that baler twine is probably the only thing holding large parts of our smallholding together!

leg problem. She seemed to be a relatively caring mum, and obviously her job was made easier by having just the one child!

Proper Fencing

It was around this time that we decided to invest in proper fencing for the pigs. We'd spent more than a year trying to get our own fencing working, and it was a mess. It had started with a core of posts and stock mesh, a steel mesh which is supposed to be stock-proof when properly installed, but the pigs tended to nose ours up out of the way, or drag it down, or rip it from the posts. We then added wooden beams to hold the bottom of the fence down, and even chipboard sections to close off bigger holes. One of the few techniques which worked on smaller holes was to tie the pieces of mesh to a post using baler twine. However, it was a losing battle, and finally we admitted defeat and decided to pay someone to put in a proper set of pens for us.

THE SHEEP FLOCK PROSPERS

While the fortunes of our pigs weren't entirely smooth, our flock of sheep prospered as the summer progressed. We had a total of forty sheep, and they did their best to keep the pasture in military close-cropped shape. This meant we didn't have to feed them. However, we did in fact choose to do so just once a day, so that we could get a good look at them, and make sure they remained accustomed to us. It was generally very easy to look after them as there was very little do, except for a small number of maintenance tasks. The most important of these was to spray them with Crovect to protect them from the flies – and it was in the accomplishing of this that we discovered that catching forty sheep presented much more than four times the effort of ten or twelve!

We had two distinct issues: one of capture and the other of control. There were no problems at all in catching the mummy Mules and Suffolks, they'd come straight to the food and into any hurdle enclosure we built. They also wouldn't get too upset if we closed the gate on them. Fortunately Muga would also always come into the enclosure, though I'm not sure why, given the behaviour of the rest of the Soays. The Soay ewes and lambs wouldn't come near the hurdles because they just didn't trust us not to do something – like spray them with bright blue liquid. Muga's other lambs, the Mule and Suffolk crosses, were also not interested, in part because of the trust issues, and in part because they'd not yet tasted the delights of ruminant mix! As much as I put out there, it was rare that any of the mummies would leave space for a lamb to get to the mix while they were eating, let alone leave any for the lambs to get to afterwards. This meant that over the summer even though we Crovected several times, we only got the Mules, Suffolks and Muga. While this was a little worrying, we figured that given the Soays' reputation for being resistant to fly strike, they would be all right. They had short coats, as did the lambs, even the crosses, and we knew this would make them less attractive to the blowflies. So we thought they'd probably be fine, we just needed to keep an extra eye on them.

It was during this time that we discovered a way of catching a single recalcitrant sheep if it was really necessary. We'd noticed one of the Mule cross lambs was limping, and fearing a return of foot rot, we were keen to take a closer look. We tried our usual repertoire of food-related tricks, but she was having none of it. We thought we might creep up on her, and as she was limping quite heavily we should be able to grab her easily. She was one of those who'd helpfully grown horns (she was about four months old at this point) – but alas, she could still move fast when she needed to, and demonstrated this whenever we got too close.

Finally we decided to herd her into a corner and try and grab her by the horns. This was a technique similar to the one we used on our first set of lambs, which always provided us with a lot of exercise, but failed every time. Personally, I think that sheep chasing could be a useful exercise in a number of sports, especially rugby, in that it consists of bursts of speed, followed by cautious walking or jogging, and an opponent that jinks about wildly. Trust me, every rugby team should have a sheep flock associated with it. . .

We were fully prepared mentally for chasing the lamb until we'd caught her. We were conscious that it would put her under a bit of stress but we did need to check her out, and we figured she would be tough enough to cope. Once again the spectacle of two people running around the fields waving their arms wildly was visible to any who were passing by and cared to look. After twenty minutes or so I came within mere inches of catching her, but missed and ended up lying in a heap on the ground, sweating profusely. Alex, who has always been much fitter than me, came across to see if I was OK. Well actually, she told me to get up, stop being a Big Girl's Blouse and sort myself out. Up I got again and wearily walked into position, another chase ensued and once again I nearly caught the lamb. It was clear she was tiring too, and now the challenge was who might collapse first.

This was now something which we had to see to the bitter end. Another miss and then, amazingly, I caught her! I achieved this major success by almost throwing myself on her, rugby-tackle style (another plus point for the rugby training, and no chance of high tackles!). She repaid my tackle by slamming her head back into my face. Remember how useful horns were for control? Suddenly they didn't seem such a great idea, as my cheek and forehead were stinging from where she'd caught me with them. By quickly moving around, I managed to move myself out of harm's way, and then just lay there for a minute or two, with the exhausted lamb clasped tightly, to recover my breath.

Alex sensibly used this time to get the hoof clippers and associated sprays and potions, so that we could treat whatever was wrong with the lamb. In my determination to catch her I'd quite forgotten what we were doing it for, so it was good that Alex was keeping her eye on the ball! Fortunately the problem was only a mild case of foot rot, and some quick clipping and a spray with the antibiotic spray soon sorted that out. I gingerly released hold of her, and she shook herself, peered at me accusingly, and then trotted off casually, as if to say, I let you catch me that time, but don't even think about trying it again.

Weaning

The wonderful summer continued and drifted slowly into a very pleasant autumn, and we continued with our daily routine, feeding the animals,

Making Hay While the Sun Shines

With all these new animals we knew that winter feed might be a little more challenging, and in an unusual example of forward planning we'd let our small field grow for hay. A local farmer who was cutting our neighbour Rob's grass for hay was happy enough to do our three acres at the same time as the hundred or so he was doing for Rob. We were hopeful for a decent harvest of a few dozen bales, enough to supply us for the winter, and perhaps even some to sell to offset the cost of making the hay! It would also be nice to be using our own hay for the first time. The weather that spring and summer in our area was perfect for hay, while other parts of the country had suffered rains and heavy floods, ruining many a hay crop. The rumours were that hay would hit record prices over the winter, and we were very glad that for once we had been prepared!

Our grass matured nicely, and as July headed towards August we suddenly became intense weather watchers. There had been a stretch of six days of sunshine early in the month, but we'd missed that, and only half of Rob's fields had been done. Now we were desperate for five days without rain, and preferably with sun, as the longer the grass stayed in the field uncut, the less value it would have as hay (the nutrients fall away once the grass has seeded); also the next growth was coming through, which would make cutting and drying it a harder process.

Then after a couple of weeks of changeable weather we were blessed with clear skies and a positive forecast for the next week. Suddenly all over the county farmers were rushing out to mow their meadows, and 'making hay while the sun shines' suddenly became real to me in a way it never had before.

We were a part of all this activity and our grass was cut, and then turned each day to dry it out. On the fourth day it was decided that it would be dry enough on the morrow for baling. This was excellent news as the weather was suddenly looking a little more forbidding, with suggestions of rain only a couple of days away. We left that morning with a sense of hope, and returned home that evening to a field full of hay bales! This was great, except given the threat of rain we knew we really needed to get all the bales in under cover, which was going to be a considerable effort. Fortunately Rob and his family offered to join us in moving them into the barn, and with much excitement we all got stuck in. We had exactly 120 bales of hay off the field, which filled almost half our main barn.

This was a definite success, as we now had enough for the winter and maybe more! This also meant we could let the animals into the small field, and give our other main field a rest. The sheep especially liked the new field, because even with the hay grass taken off it, there was still a nice layer of lovely green new growth.

checking they were all OK, and then standing back and watching them just be. All was well. Then it occurred to me that at some point we'd want to separate the lambs from their mummies and Muga, for a couple of reasons: first, because it would make them easier to control, and some would soon need to go 'on holiday'; and second, because we did not want Muga getting friendly with his daughters.

We knew it would be easier to catch the mummies, and they were now roaming over our home field (where the goats lived) and the small field. We decided to round them up and leave them in the small field, and keep the lambs in the home field. This went very well, and we even managed to move the Soays; in short, we would have considered it a great success except we failed to move Muga. He suddenly became a lot more cautious with us, and it was nearly two weeks before we could catch him.

The usual technique with Muga was to feed him, and then grab his horns and pick him up and carry him where we wanted him to go. Suffice it to say he did not enjoy the experience. Alex was worried we might have undermined him in front of the others, given this method of transport was so demeaning, but I suspect he'd have given short shrift to any ewes getting uppity after this. Despite this delay I decided that because it was quite early in the season it was unlikely that he would have had a chance to get to his daughters, so all would be well.

An Attack of Fly Strike

In the early autumn I went away again for a week, this time to sample the fine wines of Spain, leaving Alex to work and cope with the animals. With both of us looking after everything never seems to be a problem, but for only one person it becomes hard work pretty quickly.

Then one morning when Alex was working from home, she went out to feed the animals relatively early and noticed one of the Suffolks did not come to feed. Not only that, but when she spotted the ewe, it was lying down and not moving, which is a very bad sign. Alex ran over to her, to discover she was rolling her eyes in pain. Checking her to find out what was wrong, Alex found a maggot on her lower back, then lifting up the fleece a bit she realized that the whole lower part of the ewe's back was crawling with maggots – wherever she looked there were maggots. The ewe had been hit by fly strike, and badly.

Alex is always good in a crisis, and she quickly ran back and grabbed the Crovect and the clippers, the Crovect to kill the maggots and treat the strike, and the clippers to trim away the fleece to expose the affected area. Returning to the ewe she poured Crovect all over the area which had been struck, and the maggots started to boil out in a continuous stream. Alex trimmed away the fleece, some of which just fell away because of the damage the maggots had done. Realizing that the Crovect hadn't gone deep enough, she had to cut away some of the skin to reveal the worst of the fly strike. She poured even more Crovect on, and the maggots continued to come out. She held the ewe's head and stroked her, trying to calm her, and watched until the last of the maggots seemed to have dropped off dead. The ewe was still clearly in a lot of pain, but she was hopeful that she might have treated her in time. Hoping to be able to get some advice, and maybe something to treat the pain, she decided to call the vet.

The vet came out fairly quickly, as I think they treat fly strike with some priority. He checked the ewe all over, then shook his head and said the kindest option was to put the ewe down as the attack had done so much damage to her that she was unlikely to recover. He also considered she had probably been sick anyway, which is why she'd been struck so badly by the flies. This was heartbreaking for Alex. She cried when the vet put down the poor ewe, and he helpfully told her that she needed to toughen up.

In a nice way, I think his point was much as I've mentioned earlier, that when you're dealing with livestock, sometimes they just die, no matter what one does. However, I also think it's good that we still get upset when we lose an animal – I'd hate for us to become desensitized to it. I was upset that I wasn't there to support Alex during this, though I have to admit that my concern was somewhat mitigated at the thought that I hadn't had to deal with the maggots! The vet also expressed his surprise that it was the first time we'd lost a sheep to fly strike, and said we'd been lucky.

So Alex was left with a dead sheep which needed to be disposed of. At the vet's suggestion, she rang the Hampshire Hunt (HH) which, like hunts across the country, runs a disposal service for dead farm animals. Unfortunately they couldn't come that day, but would collect the ewe the following day, so Alex resolved to move the body out of the field. With her parents' help she managed to get the surprisingly heavy creature into a wheelbarrow, but unfortunately rigor mortis had set in, which meant the sheep's legs were sticking straight up, and this caused the wheelbarrow to topple from side to side alarmingly. On a few occasions it fell over, so they had to pick up the body again and get it into the barrow. The ewe had died in the middle of the main field, so it was quite a long way to the main gate, where Alex wanted to leave it to make it easier for HH to collect. The whole episode of losing the ewe in the first place, and the stressful operation of moving the body, was one that Alex definitely doesn't want to repeat. The HH also wanted to know what the ewe had been given to put her down so they could determine whether or not the carcase could be fed to the hounds, but Alex had no idea, which was awkward for all concerned. Still, they came and collected the ewe, and the incident was almost over.

The final physical effect of the disastrous day was due to the Crovect. While waiting for the vet to arrive Alex had been trying to comfort the poor ewe and had managed to get some Crovect on her face, which then had a burning sensation for the next few days! She'd also got some on her hands as she'd been picking out the maggots without using gloves, but they didn't burn – so perhaps her claim of having asbestos hands is true?

The Holiday Trip for the Lambs

The sheep were generally easy to handle. After being separated from the mummy flock, the lambs became tamer as they actually had a chance to get to the ruminant mix at feeding time, and very soon they were pushing me out of the way in their quest for the feed, which in part justified our decision to separate them from their mothers.

As the end of November approached I decided it was time to take some of them on holiday, and I booked them rooms at The Abattoir for early December. The weekend before the assigned date we'd manage to catch twelve of the lambs to tag and crutch them in preparation for their holiday. My plan was to take the girl lambs on holiday, and keep the castrated boys a little longer to get more meat out of them. This would be one of the few times boys might outlive girls on the farm! Of the twelve we caught, six were boys. We had never actually kept track of boys and girls when the lambs were born, so didn't know the ratio, and as it's usually 60:40 in favour of one gender it is useful to know!

In the course of this operation we discovered that I wasn't very good at castration – two of the boys had been only partially castrated as each retained a single testicle, and one of them had not been castrated at all! Alarm bells should probably have rung at this point, but I was clear in my plan, the girls would be going on holiday, and on the day of transport I would try and catch more of them, but at least six would go and these were the ones I tagged. I crutched all of them as it's good practice anyway.

The day before the lambs were booked in I telephoned the abattoir, to be told that I needed to bring the lambs in that day in order to hit my slot – they'd changed the rules again! I hastily secured permission to leave early from my boss and rushed home. But as the train left London it started snowing, and by the time I got back to my home station over two inches had fallen. It took me twenty-five minutes to drive home, as opposed to the normal seven, as we all had to go really slowly just to stay on the road. By the time I got home I realized two things: firstly, even if all was going well I'd miss my slot at the abattoir, and secondly there was absolutely no chance of getting there in such weather. The lambs were therefore denied their holiday for a little longer. It also gave me time to realize that the girl lambs were probably pregnant, with either Muga or one of their half-brothers the father. Therefore I didn't rebook their slot as we would need to keep them, and have the joy of some extra lambs in the spring.

As Christmas approached we moved into our winter feeding and maintenance pattern, which meant feeding everything twice a day. As the nights closed in, thoughts of sheep holidays were replaced by those of the Christmas goose. I love goose, and we'd ordered one from Waitrose specially, though at some stage I thought we might grow our own. All seemed good at this time, although a small black cloud took the edge off our midwinter celebrations: one of the ewe lambs died. A walker told our neighbours they'd seen a dead lamb in the old pig shed in the middle of the field, and they popped round to let us know. We assumed the walker was confusing a sleeping lamb, but went to check anyway. As we walked into the shed there she was, lying peacefully on the ground, looking as if she would jump up any minute to head off a-frolicking with her friends, but undoubtedly dead. There was no indication as to why she had died, it looked as if she'd just gone to sleep one night and never woken up. Still, we had twenty-two of the wonderful creatures left, and we took some joy in this as we approached the New Year.

The holiday season had well and truly finished before I once more considered taking the lambs away, though I didn't rearrange a slot until towards the end of January. Once again the Sunday before I rounded up a selection of lambs (in the sense that some allowed themselves to be rounded up!), crutched them all, and tagged the boys, including two of those with only one testicle (the intact male was much more circumspect on this occasion and had escaped the careful selection process). I was all prepared to take six of them on holiday. The day before our allocated slot I managed to leave work relatively early, with the intention of getting home and giving the lambs a quick once-over while they were eating (it would have to be fairly quick as it was both cold and dark). But once again the snow started to fall, and by the time I got home an inch had fallen, and it hadn't stopped.

I'd arranged to work from home the following morning so that I could take the lambs to the abattoir, but in the event I need not have bothered, as the snow shut down our train line completely and there was no chance of my getting into work. Sadly there was no chance of getting to the abattoir, either! Once again the holiday trip was postponed by inclement weather...

The snows that year were famously amongst the heaviest seen in the country for many a decade, and as previously described, our mornings and evenings became a slog of getting water to the various drinkers out in the fields. However, for the week that we were obliged to work from home due to the snow, we were also able to enjoy a wonderful landscape, and watching the sheep and pigs play in the snow was quite special. What was amazing was that they created the same paths in the snow as they did in the normal pasture, snow canyons along which they walked in a file to and from the water, their favoured resting areas, and the other places they roamed between. Probably this made moving around much easier for them, because after the first few trips using the tracks meant their fleeces didn't get so wet; it also meant that some parts of the field looked pristine, as if cordoned off by some invisible barrier only the animals could see. As is the way of things, the snow didn't last for ever and soon we were heading into spring.

I was becoming increasingly concerned that we had not yet taken any of our lambs on holiday, and it would not be long before the next set arrived! So once again I booked them rooms at The Abattoir, prepared the lambs, and lo and behold, all was well meteorologically and I managed to get them to their holiday home. Two weeks later we picked up our meat. This was a very special moment for us, as it was the first meat which had been born and raised at Shaw Barn, and so in a very real way was all ours: we knew exactly what the lambs had been eating for the whole of their lives, what medicines they'd had, and how they'd been treated. This may have added some extra flavour perhaps, or maybe we are just biased, but the lamb itself was glorious. The roasts were lovely, lean but full of flavour, and the steaks were the same!

Marketing Our Home-produced Lamb

We took eight lambs on holiday that time, and it was clear that we could not eat them all. We kept the three smallest for ourselves (two were pure-bred Soays, and the meat was particularly flavoursome) and I sold the rest to colleagues at work. This was a good strategy, as they were all excited to be eating our lamb, and also relatively insensitive to price as they worked in the City! I kept it reasonable, but insisted on selling in half lamb blocks, which most people probably wouldn't want. Also most of them assumed it was organic. After a while I gave up telling them it wasn't, as they really couldn't understand *why* it wasn't. The main reasons were that we didn't use organic food, and we didn't have the certification – perhaps something the organic people need to think about for clarifying their message. So every other day for a few weeks I'd take a bag of meat to work with me. I always worried that it would break open on the tube and everyone would stare at me, wondering why someone in a suit was carrying a large amount of meat around. Fortunately this never happened, and I managed to sell and deliver all the other lambs quite easily.

We were still left with a few male lambs – I thought there might have been six more (based on incomplete counts when we'd caught them). I also discovered we still had two intact male lambs, the one I'd spotted earlier and a Soay which I had never managed to catch, but noticed his assets while feeding them all one morning. This did not seem to be much of a problem until about a month later, when I noticed that Muga was becoming quite a lot more aggressive towards us. He even rammed Sue several times in one go, and we're not sure why as she didn't have any food, but had just gone into the field to sort out some bricks. My theory was that the other males were approaching maturity, and the testosterone in the air forewarning of some dominance games was getting him riled. There was only one solution, and that was that all the remaining male lambs needed to go on holiday.

Alex Takes on the Holiday Trip

Around this time Alex's work patterns had changed and she was now working from home, part-time. This was a tremendous boon for her as it gave her much more time to look after the animals, keep on top of the barn conversion, and do the prep work for her new business. This was part of her plan, the other part being to get more sleep and relax – a very fine plan I thought! I decided that I should get some benefit from the new arrangement, and so persuaded her, through the use of a dual-pronged attack of nagging and guilt, that she should take on the responsibility for the delivery of this lot of lambs to their holiday home – or the 'death process', as she likes to call it. My reasoning was that it usually required me to take a half-day holiday, and this didn't seem very reasonable. I did offer to help prep the lambs, and even load them on the trailer in the morning before I went to work, and after weeks of whinging and harassing on my part she finally caved in.

The remaining lambs had become a little calmer and more trusting since we'd taken their brethren away, possibly because they had better access to the feed now as well, and they'd definitely developed the taste for it. Whatever the reason, catching all the remaining boys was a surprisingly quick task. Alex and I took turns to crutch those who needed it, and ensure they all had tags, then we loaded them into the trailer and Alex drove off. I knew she was a little upset, but I felt it was a fair exchange, and after all, she was happy to eat the meat (though she has threatened vegetarianism on occasion since!). I then rushed back into the house to get ready for work – but as I was about to leave I realized that Alex had taken both sets of car keys, so I couldn't drive the other car to the station. So I ended up working from home until she returned – my boss was not best pleased! Still, we now knew that either of us could handle the death/holiday process if we needed to.

The year was progressing well, and we knew we'd soon have some more lambs, but we were confident we could handle anything the sheep threw at us. The thought of getting more creatures was not at all scary, and led us to adding the last members of our menagerie: the ducks and geese.

A FULL SET OF POULTRY

We had reached some form of tipping point. We now had some of virtually everything found on a normal farm, and I decided that it was time to complete the agricultural picture and go for my childhood dream of a full set of poultry. We'd managed to look after the end-to-end process of providing piglets, with mostly reasonable success, and we were getting ready to welcome our second batch of new lambs. Overall we had built up a lot of confidence in our abilities. It was also a slightly counter-intuitive way of drawing a line under any further expansion, because once you have a few of everything you don't need any more!

Finally, one of the most important lessons we had learnt was that we needed to be prepared for new animals, as opposed to getting them and stumbling our way through the first few weeks of looking after them. Though this had mostly served us well (and I think we were often harsh on ourselves), it did add an extra element of stress which we felt we could probably do without!

I already had some books on ducks and geese, picked up in earlier years when brief bursts of enthusiasm had led me to consider them, and one of the clear requirements for them was water, preferably a pond. We allegedly had a pond at the edge of our land, but it looked as if it had dried out a long time before we bought the barn, and it would have taken a great deal of work to make it viable again. It had been fed by a drain which led into the main courtyard of the barn, and when the farm had been a fully working concern, water must have flowed fairly regularly down it – but those days were long gone. It was also too far away for us to be able to access it easily, and it was right next to the woods (causing fox worries) and the footpath (causing two-legged fox worries). So I resolved to dig out a new pond.

Digging Out the New Pond

My initial plan was just to dig a big hole and let rainwater fill it, on the basis that we often had standing water and therefore the ground must be relatively impervious to water. This plan had a couple of flaws, however, one of which was that the ground wasn't really that watertight as it had plenty of fissures in it. The second was that much of the underlying base rock was chalk, which would certainly have a muddying effect on a large body of water, and that was not really what I was looking for.

Realizing I had to line it properly, I turned to that reservoir of all knowledge, the internet. It was actually fairly simple to find the basics, and before long I had ordered a large plastic lining. I decided not to buy any felt, used to protect the plastic lining from sharp bits of stone and rock when the water starts to press down as the pond fills, but instead used some old carpet from my in-laws, which saved me a lot of money!

The target area for the new pond was the half of the orchard not occupied by the pigs – where we had kept our original lambs, if only for a short time. I measured out the largest area I could between the trees, allowing them a decent area to grow their roots, and then hired a digger. A happy half-day or so later I had a half-dug-out pond. Until that point I hadn't truly appreciated just how slow digging could be, but over lunch resolved to be a bit more aggressive with my digging, and quickly got to work.

I was happily digging away, thinking to myself that I'd definitely have it finished before the end of the day, when suddenly, with the next scrape of the digger, I heard a sort of metallic snapping sound. The next thing I saw was a fountain of water starting to fill in my new hole: it looked as if I'd found a buried water pipe. I wasn't aware of any in that part of the orchard, and anyway had thought that if there were, they were likely to be shut off. While I was musing on this turn of events the water kept pumping out and was starting to turn the hole into a muddy quagmire. I realized I had no idea how to shut the water off, so swiftly headed indoors to find the bits required to seal off the flow. The best thing I could find was an old tap, still connected to a bit of pipe and a joint. Fighting the flow of water, and getting drenched as well as muddy in the process, I finally managed to get the pipes connected. My hole now had a couple of inches of water in it, and I was a complete mess – but on the upside I could wash my hands with my new tap!

I finished digging out the hole, hoping I didn't hit any more pipes, and while I did so, tried to work out where this mystery pipe had come from, and was going to. I determined that it probably went far across the fields into another field across the road owned by the farmer who'd originally sold us the place. Once the hole was finished I found the appropriate plumbing joining bits and reconnected the pipe, getting muddy again in the process – but at least it was all back as it had been. I'm fairly certain the farmer never noticed the several hour hiatus in his water supplies!

I had decided that the pond needed to be more interesting than just a shallow bowl, and so had made a few humps and bumps in it: these made perfect sense when I was planning and digging, but turned the process of laying the pond liner into more of a challenge. The carpet went down easily, but the plastic lining kept crinkling up and refused to allow itself to be flattened. But eventually I had it all settled, and held the lining down around the edges with some rocks and half bricks which had been lying around the orchard. I then filled the pond up!

I had decided that I would use the water pipes which supplied the animals' drinkers to provide the pond water, as these were nearby. This meant that

I had to disconnect the drinkers for a few hours while I redirected the water to the pond, after which I had to reconnect the pipes and make sure they were properly sealed, each time turning the whole supply off and trudging back and forth between the pipes and the taps. Clearly it would have been much easier to use a hosepipe, which is what my far more sensible wife decided to do when it was her turn to fill up the pond!

Stocking the Pond

Now we had a lovely pond, but although it was filled with water, it did seem rather empty of any interest. To resolve this issue I determined it needed some life in it, and hurried out to buy some fish. Our nearest pet store sold tropical and coldwater fish, and I headed there in a state of some excitement. I elected to buy some of the cheaper variants of the Japanese koi, because they are aesthetically pleasing, and are also supposed to be quite hardy. I added to this a couple of pond plants, though the shop's selection was rather sad, which thankfully prevented me giving them even more of my money.

When I got home I put the bags of fish into the pond. I didn't let the fish out, just put their bags in the water, where they gently bobbed; this allows the water in the bags to adjust to the same temperature as the pond water and avoids shocking the fish, which can kill them. After a couple of hours I let the little fish out with a sense of excitement. Sadly they didn't appear to share my excitement but instead seemed rather nervous. I thought this might be because there was no cover for them, so put a piece of sewer pipe connector in the pond, weighed down with a lump of chalk, and they all decided to hide in there.

After that point I rarely saw them, even when throwing their food out on to the water, unless I poked the sewer pipe with a stick. I tried not to do this too often to avoid upsetting them, but I also wanted to know if they were still alive! In an effort to get them to move around a bit more I added some more plants to the pond (purchased on the internet this time, and sent through the post), and also, from the same place, some snails. I liked the snails, and thought they might add a certain *je ne sais quoi* to the whole pond concept.

Some may wonder why I was bothering with fish and plants if the plan was to get ducks and geese (though in my defence one of the plants was duckweed, allegedly the duck version of catnip). Well, I'd read that fish would eat duck and goose poo, and remembering the whole integrated farming concept, I thought it made excellent sense, as the fish would help to keep the pond clear.

All this work had taken quite a while, and winter was fast approaching, so I decided that it was best not to get the ducks and geese until the spring, to avoid any potential settling-in problems related to cold or snowy conditions.

More Fencing Issues

We also needed time to get the fencing set up. Our original intention was to pay someone to erect fox-proof fencing all round the area, but the quotes we received were just too high, so we (Alex somewhat grudgingly) decided to do it ourselves. We bought some ten foot posts, all the mesh we would need, and prepared to erect the fence. After I'd dug the pond, Gordon had dug a trench all round the orchard area for us, so we could put the mesh into the ground to prevent any foxes getting in by digging under the fence. The trench only went down about two feet, as opposed to the recommended three, but this was because it hit the underlying chalk bedrock, which in most places was as hard as concrete, and was, after all, what our barn was built on (and in some cases built with – we discovered the first layer of bricks under the cowshed to be chalk when we were digging out the old concrete!). We figured that it would be a pretty amazing fox if it could burrow through that.

One weekend, early in the winter, we finally got to the stage of actually putting in the fence. Part of the reason for waiting was to allow the autumn and winter rains to soften up the ground for the posts, as the biggest challenge would be hammering the long posts into the ground – in part because neither of us was tall enough to reach the rammer when it was on them! We came up with a technique of using a footstool to give us the appropriate height... It was really hard work ramming the posts in with just our hand post rammer, especially from our slightly precarious perch on top of the stool. The poles were big and unwieldy and seemed to fight us, but we stuck to our task and after two weekends of effort we had them all in.

We then put mesh wire all around them. We had to do this twice as we needed two layers, one above the other, to cover the full height of the fence. The final task at this point was to tie together the two sheets of mesh between each set of posts, otherwise there would be an easy access hole for any wandering fox; for this we used the spare wires which come with the fencing mesh. And *voilà*, we had a basic eight foot (more or less) fence! It wasn't yet fox-proof as we didn't have electric or barbed wire strands sticking out a foot at the top of the fence – foxes can jump and climb. We hadn't in fact yet got the tools to construct this final component of the fence, but we did put some loops of barbed wire along the side of the fence next to the pigs, figuring that the other three sides were protected by bordering on to the field with the alpacas in it. And so we were ready!

Or so we thought ... but the pigs had other ideas! The fencing on their side had not been set into a trench as there was already a drop due to them turning everything over and snuffling along the edges of their world. We had been fairly confident it would still do the job, but as usual when it comes to pigs and fencing, we were wrong. They managed to lever up a section of the fence, and then were off for a frolic in the planned poultry portion of the orchard! They nosed around a bit and dug up a bit of the grass (of which there was very little left in their part of the orchard) – and then they found the pond. I think they thought that we'd provided this pool for their exclusive entertainment, and went straight in!

They made a right mess of it all: there was water everywhere, the sides of the pond were pulled down and half the water leaked out, and I was rather concerned that they might have poked holes through the lining with their sharp trotters. We managed to get them out and quickly fixed the fence, and checked, and strengthened, all the other potential breakage areas. We then stretched out the plastic pond liner again, replaced the stones which had been holding it down, and refilled it, crossing our fingers! Fortunately there seemed to be no holes in the lining, and it filled up nicely. I was slightly concerned about the fish, especially given how shy they were. We didn't see any for a few days, and decided to restock, so off we went to the pet shop to give them a little more of our money...

Coping with Freezing Conditions

It was a very cold winter, and the pond spent several weeks frozen over. The authorities conflict on how bad this is for pond life, but in general agree that fish need oxygen in the water, and the ice forms an effective plug and can starve the whole pond, and potentially kill all the fish and the plants. The best solution seemed to be a floating pond heater. Once more eBay provided, and within a couple of days I had a pond heater including a floater to keep it at the top of the water, some extension cables, and some boxy covers to protect the joins between the extension cables from the weather.

In the meantime I'd been breaking up the ice with a stick, and occasionally by stepping on it in the shallow end. This, I discovered shortly afterwards, is supposed to be very bad, as the crack of the ice can upset and shock the fish, and frightened fish often turn into dead fish. But unfortunately the pond heater didn't work, it just froze in place. I checked the instructions (put it in the water and plug it in), and I checked to see if there were any switches or suchlike on the heater, but there was nothing. I had clearly bought a dud. These things happen, I told myself, so I went to the garden centre and bought another one.

With this second one I read the instructions, a whole page of A4, thoroughly before doing anything. I then carefully placed it into the water before turning it on, and then flicked the switch. An hour later I checked it and it was still cold. I left it overnight and it, too, froze into the pond. I thought to myself that perhaps the cabling wasn't working, so I plugged in the electric drill, and it worked perfectly. Two strikes, I thought, and was loath to buy another heater. Then I remembered that a couple of years earlier I had in fact bought a pond heater with the plan of putting it into the main water trough to keep it ice free, but never got round to setting it up. Once again I carefully checked everything, and read the now repetitively simple instructions. Gingerly lowering it into the water I made sure the top of the heater didn't get wet, and turned it on. I left it for a while and came back hoping to feel some warmth. Nothing: it was still stone cold, and once again it froze into the pond overnight. Strike three, I thought, and decided it was clearly a sign and gave up on the pond heater for the time being.

One of the suggestions from a colleague at work was to use a tennis ball – this apparently freezes into the ice but provides an airway, so I threw in an old tennis ball we'd found in the field on one of our walks. This, combined with an occasional careful breaking of the ice, remained our strategy for the rest of the winter. It seemed to work because when the ice finally thawed we'd catch occasional glimpses of the fish, swimming in and out of their sewer pipe, and under the small amounts of water-bound foliage.

The Chickens Move In!

We moved the chickens into the new poultry area using the simple expedient of transferring their ark and shooing them along. We rather optimistically decided they would be allowed to roam free, and wouldn't need their old run as they were now in a fox-proof area. This was also because the run had finally decided to fall apart completely!

That first night when we went to put the chickens in, one was missing. We looked everywhere and couldn't find her, and in the end gave up as night had fallen and we couldn't see anything. We knew that in all likelihood she would die, probably eaten by a fox, or an owl, or some other nocturnal predator. But once again we were proved wrong, as the next morning she was still very much alive, on her feet and pecking around happily. Unfortunately she was in the old field where the ark had been, not in her nice new area, but after a little discussion and some flapping of wings we managed to persuade her to rejoin her fellows. This behaviour was repeated for a couple of nights before she finally settled in the new pen, her repeated attempts to get back to her original living area somewhat reminiscent of our early experiences with resettling the goats!

More Eggs from eBay!

As the next spring approached we still hadn't bought any new water poultry. In part I think we were a little nervous at getting some adult geese, given their reputation for being aggressive, and partly it was post-winter fatigue, which meant we were enjoying a bit more sleep and some weekends which didn't revolve around the animals. Then suddenly I had a brainwave: we would hatch out the ducks and geese ourselves! Well, not actually ourselves, but we'd get the bantams to do the hard work. Bantam chickens are renowned as excellent sitters, and we'd already had experience of using them to hatch out chicks, so how hard could ducks and geese be? This wasn't completely off the wall, because in several of the books I'd read it did mention using bantam hens to hatch duck eggs.

The next question was, where would we get the eggs from? The answer was obvious – eBay. You can go to eBay at any time and search for duck eggs, and you'll find a couple of dozen entries, and as many as a dozen or so for goose eggs. Prices vary wildly, based on the perceived aesthetic value of the waterfowl or their rarity – I've seen goose eggs go for hundreds of pounds for four, or only a few pounds each (we were definitely looking for those in the latter category).

Before we purchased the eggs we needed one of our hens to go broody; luckily we only had to wait for a short time, and then two of them went broody at the same time! We left them for a couple of days to be sure they weren't faking, and then put our bids in for four goose eggs and six duck eggs which were in the final minutes of bidding. Winning both, we paid immediately, and then waited with (almost) bated breath for the eggs to arrive; I was almost as excited as when we bought our first chicken eggs from eBay.

The eggs arrived two days later, and they were huge! The duck eggs were four times the size of normal bantam eggs, and the goose eggs were almost as big as the hens we were expecting to sit on them! There was some discussion as to how optimistic I'd been in ordering two sets of eggs and trying to do it all at once, and then, after the required several hour wait to allow the eggs to settle a bit, we went to put them under the hens.

We were very fortunate in that a third hen had gone broody – the three of them almost filled the nesting box. We decided to put four duck eggs and three goose eggs under them. However, we couldn't just leave the eggs under the hens, they needed some extra care. Eggs are turned regularly by their mothers, but as the goose eggs, in particular, were so much bigger than the hens we needed to turn them, which we did twice a day, also turning the duck eggs for good measure. In addition, as geese spend a lot of time in water their eggs regularly get covered with some dampness, and it has been found that they actually require this wetness to develop, so we had to wet the eggs each day as well. I don't think the hens were very keen on this particular procedure, but they put up with it in good humour, only pecking me occasionally to remind me to show them respect!

About a week into the process I decided that we could fit the fourth goose egg under the hens as well (my excitement just became too much for me), and thinking there might be some small hope of it hatching, I marked it 'Hope', in pencil, and placed it under the hens.

The Eggs Hatch Out!

The weeks of incubation passed slowly, at least where checking on the eggs was concerned, but every day we dutifully turned and wetted them – until one day when we opened up the nesting box and there was a big hole in one of the goose eggs, and a large beak sticking out of it! Over the course of the next few hours the gosling slowly emerged. It was a bright yellow ball of fluff, and was nearly as big as its mothers!

The next few days saw two more goslings born, and two ducklings. We didn't need to help any of them in any way, they cracked open their shells happily enough. Chick feed and water were provided regularly, and we continued to hope for more. But after a week two things became obvious: one was that the remaining eggs, including 'Hope', were not going to hatch, primarily because the hens were no longer sitting on them. The second was that these younglings were going to be a right handful for their mothers. The goslings were already as big as the hens, and were growing like topsy! Then one day the young birds discovered the pond, and from that point on were almost never off it – they clearly loved it!

So at this juncture all was going well with the new poultry enterprise: the ducklings were almost duck-sized, the geese were starting to grow their permanent feathers, and the mummy hens were still mothering them as much as was possible given their size. There was always a bit of a tussle for food, but I learnt to throw it out in several piles several feet apart, and that avoided all-out war. We seemed to have got things all in hand – and then disaster struck.

Disaster Strikes

One morning we came out to feed the poultry and nothing came to greet us. Looking all around we saw that the ducks and geese were on the pond, and they refused to leave it even when we went up close and offered them food. In some despair we started trying to find the chickens. At first we could only locate two of the hens and the cockerel, all hiding quietly in their ark, but searching the area further we found three tell-tale puffs of feathers, showing where our three other hens had been taken. Clearly a crafty fox had managed to get in. We found the likely hole, under a part of the fence bordering the pig orchard area, and we fixed that up and stacked rocks against it. At this time we were fighting with the fencing for the other pigs, and as part of putting in their electric fencing we put a single strand along the bottom of the fence between the two orchards, and I think this helped to discourage further attacks from this area.

The ducks and geese had survived the fox by taking to the pond, and for the next week or so they would rush to the pond whenever we entered the enclosure – they weren't going to trust anyone any more! They did eventually relax again, but were never quite as carefree as they had been before. They returned to greeting us for food at the gate, but they remained nervous, and never rediscovered their pre-fox innocent state. We bought some replacement hens, but they were definitely not treated with the same respect by the geese.

The Waterbirds Mature

The ducks matured into amazing black birds, with blue and green iridescent patterns to their feathers, truly beautiful. The geese grew into statuesque creatures, pure white, though one of them had some feathers on one wing which never seemed to grow out straight, and her whole wing is bent when she flaps it. Otherwise they were perfect specimens of goosehood. What was great was that they continued to hang out together; they clearly felt they were all one flock, though they did turn rather mean with the bantams, pushing them out of the way when we put food down and generally not showing appropriate respect to those who'd brought them into the world!

The geese grew into rather large birds, causing us to remember the words of the woman who'd sold us our first chickens, about geese and the mess they make. In some ways she was wrong, as they didn't seem to be doing much damage to the ground, and indeed were doing a good job of keeping the grass down – geese like to eat grass, and can be raised completely on grass if required. However, they did seem to poo an awful lot (although I'm sure the ducks contributed as well!); it was a sort of dark green, almost a teal colour, and while it quickly broke down on the grass, it was pretty nasty if stepped in. Once when they were still very small I picked one of them up, I think to put it to bed, and it covered my front with the most disgusting smelling green stuff I've ever experienced, it was astonishingly pungent. From then on I used to hold them a little way out from me as opposed to cradling them!

The worst impact of this was on the pond, which went from being a nice, fairly clear pond with some plants and fish and the occasional snail, to a stinking, fetid, greeny-black pool of poo. I could only hope that if any fish were still alive they were at least enjoying the bountiful supply of food – in fact we only saw fish again many months later when one was floating dead on the top of the pool. It was quite big, though, which might indicate that they were (or had been) prospering. To date we've not yet developed a satisfactory way of keeping the pondwater clean, and it continues to be rather unpleasant!

As the next winter approached I started to worry about the ducks being frozen into the pond. This is a real risk, and several of the duck books had mentioned it. Also, if you were to google 'ducks frozen in ponds' you'd see an amazing number of pictures. I have to admit it seemed odd to me that evolution had not bred this tendency out of them, but nonetheless the evidence was clear. I was just in the process of planning out a second wave of pond-heating attempts when a really cold spell struck; it was one of the coldest Novembers in fifty years, and the pond froze over. The ducks and the geese, however, were absolutely fine, and none of them came even close to being frozen in. Actually they were most annoyed that they'd lost their pond, but they were at least alive and well!

Apart from swimming, the geese also use the pond water to clean themselves, and it was this I think they missed the most (though given how mucky it was it always seemed a slightly pointless activity). So we took buckets of water out for them, and the first thing they would do was start preening themselves, using the water to bring their feathers back to the lustrous pure white which was their source of sartorial self-esteem.

The survival of the ducks despite the dangers of being frozen in the pond was very encouraging, and made me wonder if those ducks which froze had been caught for different reasons. Nonetheless it meant that during the winter we could concentrate on getting them water and food, and could safely ignore the pond.

First Steps in Goose Husbandry

Towards the end of the winter the ducks and geese were old enough and mature enough to start showing what gender they were. One of the biggest problems with hatching eggs, as opposed to buying fowl directly, is that not only do you not know how many will hatch, but you have no idea what sex the younglings will be. It's also very hard to identify their gender when they're very young, when it is easier, from a practical and ethical perspective, to send the males to the great pond in the sky (if you see what I mean). Male birds, like all male farm animals, are undesirable, and some would argue that even one is too many, so the sensible thing to do is to cull them out when they are young to avoid later complications. There are people who can 'sex' chickens, and some who claim to be able to do so with eggs; the rest of us, however, have to guess, or wait for the birds to show obvious signs – and with our geese and ducks it took a while to become clear.

The ducks seemed to be one duck and one drake, which was positive. We based this surmise on the fact that one of them was bigger and heavier than the other, and slightly more forward, and we decided that it was probably the drake – though we'd have to wait to see if one or both laid eggs to be truly sure. With the geese it was more obvious – we had two ganders and a goose. The ganders looked exactly the same as the goose, they were just much, much louder and more aggressive. The goose was almost always silent – I've heard her speak on only a couple of occasions. The ganders, however, would talk (or rather, shout) constantly at us, and also hiss! They were also aggressive with food, and until we showed them who was boss they went through a stage of trying to attack us. The most ridiculous thing would be when they were hissing at us, while eating out of our hands: they'd hiss loudly, peck the food, hiss a bit more, sometimes with the food in their beaks, peck a bit more… They clearly needed a lesson in who was top of the pecking order!

We learned a bit of goose husbandry from a friend who'd lived in the country all his life – except for a brief period slaving in a City law firm – and while we thought it sounded a bit mad, we thought we'd try it. The trick, apparently, was to pick up the loudest goose, and tucking it under one's arm, spank its bottom in front of all its friends. Now I know how I'd feel if someone did that to me, but I wasn't sure geese would be the same! Alex was the first to catch and spank one of the ganders, and it did seem to calm them down a bit – though she did say it was difficult to actually spank a goose as its bottom was soft and downy like a pillow, so it was more of an exaggerated tap! Once she'd done it a few times they became much better at evading her, but also much less likely to go for either of us, and while they still hissed there were far fewer attempts to peck the back of our legs!

We soon came to enjoy their talkative ways, and when we fed them and they were squawking and the ducks were quacking it made me realize that I really *was* living my dream about having poultry.

CODA: LIFE OF POO

THE COUNTRY IDYLL

Over nearly five years of building our menagerie we've experienced some lows, and some really amazing highs. I don't regret the adventure, and even the mistakes and mishaps taught us valuable lessons. I'd say keeping farm animals is not for the faint-hearted, and that might be fair, but I'd recommend it to anyone who can give them time and energy. At the very least everyone should have chickens! I shall leave you with some musings that I wrote nearly three years into the experiment.

I've read a lot about leaving the city and moving to the country. There is much mention of idyll, growing your own vegetables, beautiful scenery, keeping animals and suchlike, but very little mention of poo. I feel this is a shocking omission, and I hope this piece starts to rectify this gap in the literature.

My wife and I moved from Islington to Hampshire nearly three years ago, and while she had some country experience, I had been a suburbanite/city dweller all my life. This didn't hold us back, and over the course of the last few years we have built up a small menagerie, including fish, a cat, dogs, goats, chickens, cows and sheep – all of which are mostly hardy animals requiring minimal attention. However, they do require some care, and it is not uncommon for me to have a 'to do' list of dealing with the following:

* Fish poo
* Dog poo
* Cat poo
* Chicken poo
* Goat poo
* Cow and sheep poo

I'll talk through each of these briefly...

Fish poo: This is easy, you just need to clear out the tank, otherwise the ammonia builds up to toxic levels. This is usually done by pumping out the water (often by siphon, I just wish I could learn to stop before the water gets into my mouth), and replacing it with new treated water. I do wonder if the used water could be useful for something, but given the other chemicals we put in to clean it I am loath to try it out as yet.

Dog poo: Our dogs have a largish area they can run around in, and obviously they often do their business in there – although I sometimes suspect them of holding it in for walks so they can pick the most embarrassing times and places to relieve themselves. We used to burn their poo, but about a year ago I got the dog poo version of the Tumbleweed worm farm (well, it's actually exactly the same as the normal version, just with an extra set of instructions). Each time I collect up their poo I put it in the worm farm, where it is slowly turned into good quality compost: the worms seem to love it. However, there are two important points to remember if you choose to follow this route: one, don't put dog poo into the worm farm for a week after worming the dogs, as this will kill, or at least damage, your worms; and two, don't put the compost on to anything you will eat, just in case (dogs have a few nasties in their poo which can transfer to humans if ingested, even though this is rare) – but roses and young oaks seem to really love it!

Cat poo: Cleaning out the cat box used to be a painful exercise, but fortunately she now goes somewhere else outside. I fear finding out where exactly, but until then this is an easy one! I had thought we could feed it to the worms as well, but that doesn't seem recommended, and I'm not quite ready to experiment.

Chicken poo: We have an ark and clean it out regularly. In general this involves putting on gloves and, with the aid of a stick, scraping and lifting out the poo and sawdust/shavings mix, and then replacing it with new shavings. This is often made more 'fun' by the goats, who like to eat the shavings, until they remember they don't like them, try and eat the plastic, and then jump into the wheelbarrow. Once collected, the chicken poo and shavings make a good compost or compost starter, but the efforts of the last couple of years ended up feeding nettles as they aggressively invaded my compost patch. Still, we have plans to get better at this. ...

Goat poo: The goats have their own hut (well, shed), which they liberally poo in. When I put straw down they insist on eating half of it, which is somewhat frustrating, but the remnants make it a bit more comfortable for them, and easier to clean out. I have several books on goats, and a couple of these claim that goats love the build-up of their faeces and straw as it creates a warm fug for them; the others say that goats much prefer very clean spaces and get upset if their area isn't cleaned up regularly. After watching them I'm not sure they actually care, but I suspect they like fresh straw (partly to snack on), even if it just goes over the existing layer. Still, I like to clean it out fairly regularly (I have to admit it's become more infrequent over the years) – more often in the winter – to avoid it getting nasty.

Cow and sheep poo: When we first started (before we had the sheep) this job involved me roaming the field looking for cowpats to put into my wheelbarrow to take to my manure pile. This was depressing, as the cows had no particular favourite place (not even the shelter which I strewed straw in and they ignored, preferring to be under the trees...), and I never seemed to amass much. Even worse, the nettles also got the results of all this effort. However, we recently laid out a large concrete area as the base for some future buildings, and that's where we feed the animals, and also where they often relieve themselves. Therefore I have a ready area to scrape up all that lovely manure, ready to let it rot down for the winter in preparation for spring planting. This does challenge us with some big piles of rotting poo in awkward places, but we try our best to work around it.

Pig poo: I hear that pigs often pick a corner of their patch and do all their business there, which sounds too easy to me. Alternatively it is said that they plough it into the field as they snuffle around, which also sounds rather unlikely! Pigs are on our list to look at, so I'm sure there'll be another entry on my 'to do' list soon enough. (This turned out not to be the case with our pigs: they tend to go pretty much anywhere...)

I hope these scatological musings are helpful and not too unpleasant; I certainly wish someone had shared such information with me in the early days. However, I fear my old city self would (after stepping back a pace or two) have been somewhat disturbed by this 'to do' list. Still, I have to say, given that dealing with poo is one of the duties of having animals, I think it's a light one, and I positively revel in it (figuratively, that is).

Index